Construction safety handbook

Second edition

Construction safety handbook

Second edition

V. J. Davies, CEng, FICE
K. Tomasin, DLC(Hons)Eng, CEng, MICE, FIExpE

 Thomas Telford

Published by Thomas Telford Publishing, Thomas Telford Services Ltd, 1 Heron Quay, London E14 4JD.

First published 1990
Second edition 1996

Distributors for Thomas Telford books are
USA: American Society of Civil Engineers, Publications Sales Department, 345 East 47th Street, New York, NY 10017–2398
Japan: Maruzen Co. Ltd, Book Department, 3–10 Nihonbashi 2 chrome, Chuo-Nku, Tokyo 103
Australia: DA Books and Journals, 648 Whitehorse Road, Mitcham 3132, Victoria
A catalogue record for this book is available from the British Library

Classification
Availability: Unrestricted
Content: Advice/guidance based on best practice and current legislation
Status: Authors' opinion
User: All construction professionals

ISBN: 0 7277 2519 X

Typeset by WestKey Ltd, Falmouth, Cornwall
Printed in Great Britain by Redwood Books, Trowbridge, Wiltshire

Foreword

'Why?' is the first reaction to any new edition of a book long recognised as comprehensive, thorough and practical; the reputation enjoyed by *Construction safety handbook*. Why at all? Why now? Why my money?

The answer is that safety management has now been given a whole-industry prominence lacking in the late 1980s. EU directives transposed into regulations have reinforced the goal-setting philosophy of the Health and Safety at Work, etc., Act 1974 (HSW Act). Impetus initially stemmed from the core obligations of the European Commission's framework directive (89/391/EEC). These are now written into daughter directives and into regulations and the like stemming from them. They are:

The principles of prevention and protection (Article 6(2))

briefly:

- if practicable, avoid risks
- evaluate risks that cannot be avoided
- combat these residual risks at source

while

- adapting work to individuals
- adapting to technical progress
- replacing the dangerous by the non-dangerous or by the less dangerous

and

- giving collective protective measures priority
- giving appropriate instructions.

The principles of risk assessment (Article 6(3))

- evaluate the hazards to the health and safety of workers
- hence evaluate the preventative methods to assure improved protection, integrated into all the activities of an organisation
- these preventative methods and tasks are to
  - take account of workers' capabilities

- ○ involve consultation
- ○ ensure only adequately instructed workers have access to areas where there is serious and specific danger.

There are important provisos. First, the depth of risk analysis and the action required should – must – be proportionate to the perceived hazards. Second, avoid bureaucracy. Records, plans, files must be pruned to their purpose – to inform, to manage, to instruct. Third, build on experience. Learn from it, as is self-evident in chapter 4 of this handbook. Safeguarding health and safety is always an important consideration. Only exceptionally will it be the only consideration.

An early daughter directive was transposed into the Management of Health and Safety at Work Regulations 1992. These, biting on all industry and all of commerce, have been absorbed into general management practice. Regulations affecting construction workplaces are part of the recent Construction (Health, Safety and Welfare) Regulations 1996. Advantage has been taken of the need to transpose the remaining annexes of the *Sites Safety Directive* (92/57/EEC) to consolidate the Construction Regulations of 1961 and 1966 in less prescriptive terms. Consequently, 90 existing provisions in the former Construction Regulations have been replaced by 30 or so new requirements giving management more responsibility for deciding what meets the criterion 'as far as is reasonably practicable'.

Important as these are, they pale in comparison with the sea change imposed by the Construction (Design and Management) Regulations 1994 (CDM). Although the general duties of the HSW Act apply to all types of work activities and situations, they implicitly assume that enterprises have coherent management that is able to learn from repeated experience. Not so with construction. Projects are normally undertaken by temporary coalitions with no continuing involvement.

Briefly, the CDM Regulations set out to:

- create a stand of health and safety management running through the disparate organisations contributing to any project
- involve to this end all participants, clients, designers, health and safety co-ordinators and contractors of every kind and self-employed persons
- ensure that relevant health and safety information is sought and passed down the chain of participants
- require designers to accept as a design consideration the adequate regard for the health and safety of persons at work and of any person who might be affected by them
- achieve the appointment (from a health and safety standpoint) of competent and adequately resourced designers, health and safety co-ordinators and contractors.

The CDM Regulations were bold (and indeed remarkable) in that a management system was imposed on a fragmented industry without the benefit of prior field trials. Necessarily they are general so that they may be adapted to the varying circumstances of projects. Those who expect the assurance of prescriptive directions will be disappointed. Good management and sound judgement will decide the appropriate response.

Many pages of this second edition make reference to the CDM Regulations. Practical safety and health management will be little affected apart from more emphasis on information passing and on co-ordination. This apart, clients and designers now have their duties, implicit in the HSW Act, made explicit with a much greater emphasis on foresight by the avoidance and anticipation of hazards.

What is needed? Nothing extraordinary:

- a working knowledge of the Regulations and sound safety practice – the bedrock of this handbook
- a common sense application of the principles of prevention and protection following risk assessment attuned to the circumstances in hand
- hence proportionate action with an auditable trace
- co-operation, co-ordination and commitment.

These are the thrust of this second edition of the *Construction safety handbook*. Throughout the edition's analysis knowledge, proportionate action, co-operation and commitment are assumed. These must form the basis of training and practice. The 'whys' are answered.

The long-term object of a safety culture will not happen quickly. Construction is too fragmented and diverse. Knowledge and commitment will shorten that time, to the advantage of all, the better management of projects especially. This handbook provides knowledge. Readers will provide commitment.

Professor Donald Bishop
March 1996

Acknowledgements

The authors wish to thank their former colleagues in the Health and Safety Executive and W. S. Atkins Consultants for permission to use material in this book based on papers and other literature prepared in the course of their employment. They also wish to thank the British Standards Institution for permission to draw from contributions to several codes of practice and to the Health and Safety Executive for information from several of its publications.

The authors greatly appreciate the valuable help given them in the preparation of the first edition by John Lomax, OBE, Safety Consultant (chapters 8 and 9); Malcolm Tucker, John Laing Construction Ltd (chapter 5); Clive Tucker, formerly of British Rail (chapter 5); Anthony Cuming, W. S. Atkins Engineering Sciences (chapter 7); Tim Foley, W. S. Atkins Consultants (chapter 2, part 2); and Mr Tomasin's daughter Jill (chapter 2).

Contents

Introduction

Attention to matters of health and safety is a responsibility of everyone at work but it is of particular importance in the construction industry where the accident rate is so high. The initiative for this book developed from discussions on safety training by the Safety Panel of the Institution of Civil Engineers. Effective training of professionals in the construction industry is one of the means by which safety management can be improved and the number of injuries and fatalities reduced. It was recognised that there was a need for a textbook for graduate civil engineers wishing to study for their professional examinations and to give qualified professional engineers, architects, surveyors and builders as part of their continuing education, sufficient information to

- take care for their own safety
- manage the safe conduct of those employees for whom they are responsible
- fulfil their moral, statutory and legal duty of care for the safety of others.

The book therefore has been written primarily by civil engineers for civil engineers but its contents are relevant for architects, surveyors, builders and all professionals and managers engaged in construction and demolition.

Responsibility for safety and health rests with all construction participants, including clients. It is not only confined to construction work on site. Design engineers, architects and surveyors are exposed to hazards during the investigatory stage of a project and while carrying out inspection tasks during the construction phase and on completed works. In addition, designers have a statutory duty, a moral responsibility and a duty of care for the safety of building workers, maintenance staff, in some cases, demolition workers, and towards the public in general.

To avoid the repeated use of the word 'person' the authors have resorted to a number of alternative words and phrases that do not mean precisely the same. Where 'a workman might be injured' is used, 'a person might be injured' is meant. The persons will usually be workmen but not always – they might be professional staff, they might be visitors to the site, or even members of the public. The masculine gender has been used for convenience in several places but it is not to imply any discrimination on the part of the authors and we sincerely trust that the increasing number of female civil engineers, architects, builders and surveyors will not be offended.

Introduction to the second edition

Since this book was first published a number of new health and safety regulations have been introduced most notably the Management of Health and Safety at Work Regulations 1992, the five associated sets of regulations of 1992 and the two important sets of regulations that only apply to construction namely the Construction (Design and Management) Regulations 1994 and the Construction (Health, Safety and Welfare) Regulations 1996. This second edition has been modified throughout to take account of these regulations and the procedures necessary to comply with them. In the intervening six years there has been a lot of activity on the occupational health and safety front and this has led to a number of other changes to the original text not connected with the major new regulations.

What have not changed to any extent are the causes of injury and death on construction sites. It is true that there seems to be a welcome decline in the number of reported injuries in recent years but accidents on site continue to happen at an unacceptable rate. As a result there has been little need to revise those sections of the book devoted to the hazards of construction and their prevention. Moreover, apart from the introduction of the planning supervisor and his duties under the CDM Regulations, the general principles of effective management of construction work as described in the first edition have required updating only to the extent that what were then recommended procedures have, since 1992 and 1994 become mandatory.

1 The problem

Safety and health

Safety in the context of civil engineering is the discipline of preserving the health of those who build, operate, maintain and demolish engineering works and of others affected by those works. Safety is defined as the freedom from danger of risks. This can apply equally to the danger of physical injury and to the risk of damage to health over a period of time. However, in this book the term safety generally applies to the freedom from risk of injury from accidents whereas health damage arises from both immediate and longer-term effects of exposure to an unhealthy working environment.

Safety

Accidents that occur during construction and demolition activities result in injury, mostly, but not invariably, to employees on the site. Accidents can occur even before works begin, during survey and investigatory phases of a project, and they can also occur after the works have been completed, because of faulty design or construction, causing death or injury to those engaged on maintenance work and to members of the public.

The construction industry accident record

In a typical decade about 1500 people are killed on construction sites in Britain and 25 000–30 000 more are seriously injured. In addition, 300 000–400 000 suffer injuries sufficient to keep them off their normal work for at least three days.

Table 1.1 shows the annual numbers of deaths and major non-fatal injuries to employees in the construction industry reported to the Health and Safety Executive for the years 1981–95. In 1985 the procedures for reporting injuries, diseases and dangerous occurrences (*see* p. 275) were revised which disturbed the continuity of major injury statistics. While there appears to be some improvement in site safety in recent years, it is known that the statistics are unreliable due to under-reporting. In spite of

Table 1.1. Fatal and major injuries in construction

	1981	1982	1983	1984	1985	1986 Jan/Mar	1986/7	1987/8	1988/9	1989/90	1990/1	1991/2	1992/3	1993/4	1994/5
Deaths	128	135	150	130	138	26	125	143	137	154	124	99	96	88	74
Major injuries	1727	1999	2232	2356	2357	576	3179	3328	3660	4107	3838	3299	2740	2567	2582
Total	1855	2134	2382	2486	2495	602	3304	3471	3797	4261	4062	3398	2866	2655	2656

Note 1. From 1986 the reporting year is April to March.
2. Fatalities and major injuries to the employed and self-employed are included in the table but not the non-employed.
3. The continuity of major injury statistics was disturbed in 1986 by a revised definition of a major injury. See p. 275.
Source: Health and Safety Executive.

Table 1.2. Fatal and major injury rates per 100 000 employees

	1981	1982	1983	1984	1985	1986/7	1987/8	1988/9	1989/90	1990/1	1991/2	1992/3	1993/4	1994/5
Fatal injury														
Construction industry	9.7	9.7	11.6	9.8	10.5	10.2	10.3	9.8	9.4	9.3	8.8	8.0	8.4	6.2
Manufacturing industry	2.0	2.4	2.2	2.7	2.4	2.1	1.9	1.8	2.1	1.8	1.5	1.5	1.6	1.1
Major injury														
Construction industry	155	188	213	225	225	282	276	282	299	286	272	239	208	211
Manufacturing industry	69	72	80	90	92	145	142	141	144	136	129	124	126	125

Note revised definition of major injury from 1986. See footnote to Table 1.1.

this apparent improvement it is clear that the injury rate for the construction industry is still at an unacceptably high level.

Unfortunately, it is not only the construction workers themselves who suffer injuries and death. In the past decade 102 people not employed in the industry, including children and old people, were killed on construction sites or because of construction activities, and a further 1251 were seriously injured.

In addition to all the injuries reported to the Health and Safety Executive, there are some 4000 road traffic accidents reported to the police every year involving one or more casualties at the site of roadworks. How many of these are due to the shortcomings of the construction industry is unknown.

To put accident figures in perspective they must be linked to the number of people at risk. Thus, comparisons of incidence rates, which are the number of fatalities and major injuries per 100 000 employed, provide a means of assessing the relative danger for people at work in various industries. Table 1.2 compares the injury rates of the manufacturing industry with those of the construction industry for 1981 to 1994–95. In the construction industry the risk of major injury is almost twice that for manufacturing and the risk of a fatal accident nearly five times greater than in the manufacturing industry.

Economic effect of accidents. The accident statistics represent not only terrible human tragedies but also substantial economic cost because accidents also cause

- damage to plant and equipment
- damage to work already completed
- loss of productive work time while debris is cleared and damaged work rebuilt
- reduced workrate until normal site working rhythm and morale are restored
- disruption while investigations are carried out by the company safety department, the insurers, inspectors from the Health and Safety Executive and sometimes representatives from the trade unions
- legal costs and, in some cases, fines
- increased insurance premiums
- loss of confidence and reputation.

All this has to be balanced against the cost of working safely. Insurers can show that the modest extra cost of a safe, orderly site is well invested because the risk and cost of serious accidents are so high.

Reasons for the poor record. There are a number of reasons why the accident record in the construction industry compares poorly with that of the manufacturing industry.

In factories there is normally a controlled working environment, with little change in working procedures and equipment for long periods, and the labour force usually remains fairly constant. Hazards, once identified, can be remedied with relative ease and the danger can be overcome: for example, a guard may be designed and fitted to an unfenced part of dangerous machinery.

However, in the construction industry the working environment is constantly changing, sites exist for a relatively short time and the activities and inherent risks change daily. Within a short time of a hazard being identified and dealt with, the work scene has changed, bringing new hazards. There is also a high turnover in the workforce, which means safety awareness is not always as good as it should be.

The complexity of the construction industry with its many small firms, subcontractors and self-employed labour militates against the establishment of safe working practices. The financial resources of small firms are generally insufficient to provide the necessary high standard of safety training carried out by many of the larger firms. Moreover, the managerial knowledge of many small contractors is based on experience and often lacking in theoretical background. Many small contractors do not employ a professional safety adviser and have neither the time nor the inclination to keep abreast of legal requirements and technical developments in safety and health matters. Compared with accidents in factories there are many more accidents in the construction industry that cause multiple fatalities and serious injuries.

When competition is particularly fierce, some contractors are prepared to ignore safety legislation and good working practice in order to bid competitively. Unfortunately, there are also clients who are willing to employ these contractors regardless of their competence in safety management.

Safety legislation and, in particular, the regulations that apply to construction (*see* p.19) provide a sound basis for good safe building practice. However, the Factory Inspectorate of the Health and Safety Executive, whose duty it is to enforce the law, does not have the resources to supervise every site. Unfortunately, therefore, law breaking is rife because the risk of prosecution is low and, moreover, the practice of cutting corners for commercial gain is frequently condoned by senior management and ignored by clients and their professional advisers.

The majority of accidents can be prevented by dedicated safety management and a disciplined approach to ensure compliance with safety rules and procedures by all involved in construction operations. This should be evident from the top management down. The tough image which flouts all foreseeable dangers is a major obstacle to be overcome for it not only causes individual casualties but also affects large sections of the industry, resulting in the widespread axiom that 'if you aren't prepared to take risks you shouldn't be in the business'.

Causes of accidents

The key to influencing safety and health is to foresee the hazards and thus be in a position to eliminate them. The main causes of accidents are

- falls
- stepping on or striking against objects
- lifting and carrying – over-exertion
- machinery
- electricity
- transport
- fires and explosions.

By far the largest category is falls, which includes people falling from one level to another, people falling at the same level and plant and material falling (including a structure or part of a structure collapsing) and striking, crushing or burying people. Each year 70–80% of all fatalities and 35–40% of all injuries may be attributable to this cause.

About 25% of all the accidents that result in workers being unable to carry out their normal work for at least three days are due to incorrect manual lifting and carrying, generally of too-heavy loads. This usually results in strain and sprain injuries. A further 10% are due to stepping on or striking against objects, for example, stepping on protruding nails left in discarded timber. Fortunately, few accidents in these categories are fatal.

Fatalities. An analysis of fatal accidents in the building and civil engineering industries published by the Health and Safety Executive in 1988 entitled *Blackspot construction* describes how people die.[1] Of the 681 deaths in the five years from 1981 to 1985, the causes were as shown in Table 1.3. Ten years later the figures are still valid.

Figure 1.1 illustrates HM Factory Inspectorate's distribution of fatalities by occupation for 1981–85. It takes no account of the numbers employed in the various occupations and trades and should, therefore, not be regarded as an indicator of the relative risk attached to each group. However, professional engineers, together with managerial staff, suffered 49 fatalities – some 7% of the total – during the period.

Table 1.3. Causes of fatal injuries in the construction industry for 1981–85

Falls (of people)	52%
Falling material/objects	19%
Transport/mobile plant	18%
Electricity	5%
Asphyxiation/drowning	3%
Fire/explosions	2%
Miscellaneous	1%

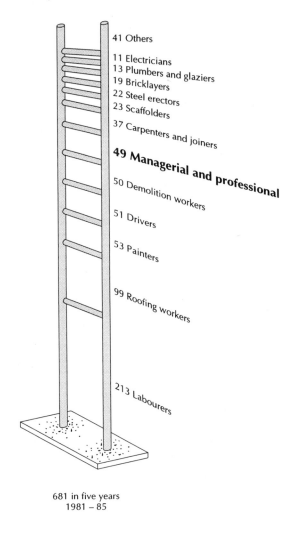

41 Others

11 Electricians
13 Plumbers and glaziers
19 Bricklayers
22 Steel erectors
23 Scaffolders

37 Carpenters and joiners

49 Managerial and professional

50 Demolition workers

51 Drivers

53 Painters

99 Roofing workers

213 Labourers

681 in five years
1981 – 85

Fig. 1.1. HM Factory Inspectorate's distribution of fatalities by occupation for 1981–85 (courtesy HSE; source *Blackspot construction*)

Major injuries. Specified major injuries are defined by the Health and Safety Executive for reporting purposes and are listed in chapter 11 (on p. 275). The construction industry hospital (Fig. 1.2) illustrates the approximate distribution of all major injuries that occur in a typical year, from which it can be seen that 73% were due to people, materials or objects falling.

Reliability of the statistics
While the number of fatalities at work reported to HM Factory Inspectorate can be expected to be reasonably accurate, there is reason to suspect

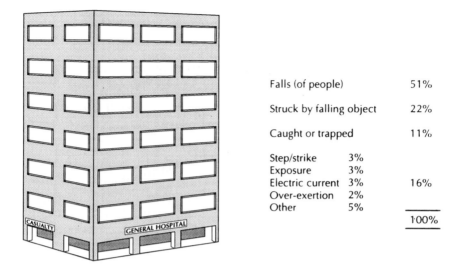

Falls (of people)		51%
Struck by falling object		22%
Caught or trapped		11%
Step/strike	3%	
Exposure	3%	
Electric current	3%	16%
Over-exertion	2%	
Other	5%	
		100%

Fig. 1.2. The construction industry hospital: causes of major injuries in a typical year (compiled from HSE sources)

under-reporting of non-fatal injuries required by the Reporting of Injuries, Diseases and Dangerous Occurrences Regulations (RIDDOR) (*see* p. 275). The reasons for this are that some firms may not comply with reporting procedures, either because of ignorance or from unwillingness to disclose accidents, and also that the growth of subcontracting and use of self-employed labour on construction sites has tended to confuse the responsibility for reporting. In addition, some people who have suffered injuries that keep them off normal work for more than three days continue work on light duties, and their injury is often wrongly not reported. It is likely therefore that the accident rate in the industry is somewhat worse than the statistics show.

However there should be an improvement in the level of reporting site injuries from 1995 because the Construction (Design and Management) Regulations 1994 (*see* p. 20) imposed duties on the principal contractor to monitor the contractor's performance, and on the contractors to keep the principal contractor informed of injuries on site.

In 1994 the Health and Safety Executive circulated a consultative document for revised regulations on the reporting of injuries, diseases and dangerous occurrences (*see* p. 275) aimed at improving the quality and intelligence of accident data. The revised regulations, the Reporting of Injuries, Diseases and Dangerous Occurrences Regulations 1995 (RIDDOR 95) came into force in April 1996.

Health

Occupational ill-health statistics

In comparison with accident statistics, the recording of cases of occupational diseases is beset with problems of two sorts

- defining what is meant by occupationally related ill health
- devising reliable methods to count (or estimate) those occurrences of ill health falling within the definition.

Nevertheless, the Health and Safety Executive publish statistics based on cases of notifiable diseases reported to them and obtained from other sources. Table 1.4 gives an annual estimate of numbers suffering from work-related ill health in the construction industry with a 95% confidence limit. The procedure for reporting diseases is described on p. 275.

Table 1.4. Annual estimate of numbers suffering from work related ill-health in the construction industry (95% confidence limit)

Hazard	Possible resulting disease or condition	lower limit	upper limit
Asbestos	Mesothelioma, asbestosis, lung cancer	200–250 deaths	
Musculoskeletal injury	Total	30 400	48 100
	Back disorders	14 700	27 800
	Upper limb disorders	4 600	13 000
	Lower limb disorders	1 100	6 500
	Unspecified	3 800	11 700
Respiratory disease	Bronchitis, emphysema	2 000	8 500
	Asthma	1 000	5 800
	Pneumoconiosis (excl. asbestosis)	300	4 400
	Sinusitis, influenza	300	4 400
Skin disease	Dermatitic conditions	3 100	10 500
Noise	Deafness, tinnitus	1 000	5 800
Ionising radiations	Radiation exposure (more than 15 mSv)	50	
Lead	Lead poisoning (above suspension level of 69 g/100 ml)	18	
Compressed air	Decompression sickness	50	

Compiled by HSE from various sources and published in full in *A guide for designers on the Construction (Design and Management) Regulations 1994.*

Health hazards. The major health problems associated with construction work may be categorised into

- chemical hazards: those due to liquids, dusts, gas and fumes
- physical hazards: those due to cold, heat, noise, vibration, ionising radiations, compressed air, lasers and manual handling.

For works in sewers and on contaminated sites there is also a risk of biological infection. The common health hazards of construction operations are summarised on p. 158 *et seq.*

Employers are required by law (*see* p. 158) to identify health hazards associated with the work of their employees and to take the necessary steps to remove the hazards or to give adequate protection both to their employees and anyone else who may be affected.

2 The law

Introduction

An examination of safety and health in the construction industry would be unrealistic and inadequate without at least a brief consideration of the law. A detailed analysis of all the legal problems likely to affect engineers wherever they are on the management ladder is not given here, but engineers should be aware of the basic nature of their duties and liabilities both in criminal and civil law. In general terms there are two main branches of English law distinguished not in the nature of the act someone commits (for the same act can constitute both a crime and a civil wrong) but in the legal consequences that may follow from that act. The terminology relating to both branches of the law differs and the criminal and civil courts are also almost entirely distinct.

PART 1 CRIMINAL LAW

Basis of criminal law

The law relating to health and safety at work is predominantly criminal law enforced through proceedings brought in the magistrates' and Crown courts. On conviction, any employee or his company may be fined a sum which is paid to the State, or in extreme cases he (or an appropriate officer of the company) may be sentenced to a term of imprisonment.

The most important form of legislation is the Act of Parliament and much of health and safety law is embodied in Statute. The Health and Safety at Work etc. Act 1974 (the HSW Act) is an example of such an Act of Parliament. There is, in addition, delegated or subordinate legislation; Government orders known as statutory instruments, which often contain detailed rules and regulations relating to Acts of Parliament. The Construction (Design and Management) Regulations 1994 (the CDM Regulations) made under the HSW Act is an example of delegated legislation.

History of construction health and safety legislation

As long ago as 1904 concern was shown over accidents and working conditions in building work. A public enquiry was held, but it was not until 1926 that the first building regulations were passed by Parliament. However, they were of limited scope and only applied to sites on which mechanical power was used. In 1931, their scope was widened to cover the growing number of cranes used in building work, the main types of which were scotch derricks and guyed derricks, where the anchorage and ballasting arrangements were often inadequate.

By the late 1930s these regulations had become outdated but because of the intervention of the Second World War, it was not until 1948 that the first comprehensive set of building regulations was compiled. They applied to all building sites (but not to civil engineering works) where persons were employed. These became law following a public enquiry and were entitled the Building (Safety Health and Welfare) Regulations 1948.

The post-war expansion of the construction industry, coupled with an equally large increase in the accident rate and the need to improve safety in the civil engineering section of the industry, called for more complex and comprehensive legislation. This resulted in the replacement of the 1948 Building Regulations by four statutory instruments collectively known as the Construction Regulations. The first two, the Construction (General Provisions) Regulations 1961 and the Construction (Lifting Operations) Regulations 1961 came into operation in 1962. Civil engineering works were then for the first time covered by these legal requirements. They were followed in 1966 by the Construction (Working Places) Regulations 1966 and the Construction (Health and Welfare) Regulations 1966. The Construction Regulations of 1961 and 1966 were statutory instruments made by various ministers of labour through powers conferred on them by the Factories Act 1937, 1948, 1959 and 1961.

The Factories Act 1961 was the principal Act concerned with the safety, health and welfare of people employed in factories, although the provisions were extended to places other than factories. Section 127 of the Factories Act stipulated the parts which applied to building operations and work of engineering construction (civil engineering work). Also, under this section, building and civil engineering operations were subject to section 76, which provided the powers for the Secretary of State for Employment to make special regulations for safety and health. Other provisions under section 127 extended the powers and duties of HM Inspectors of Factories, to cover construction work, e.g. to visit and inspect construction sites.

In 1972 a report of the Committee on Safety and Health at Work (the Robens Report) proposed many changes in the law and administration in relation to occupational health and safety. The majority of its recommendations were implemented in the HSW Act. This Act places broad general duties on employers, the self-employed, employees, manufacturers of industrial products for use at work and persons who to any extent have control of premises where people work. However it does not apply to domestic servants in private households.

The HSW Act established the Health and Safety Commission and the Health and Safety Executive and gave powers of enforcement to Health and Safety Executive inspectors.

Earlier legislation was not immediately repealed upon the introduction of the HSW Act. A number of Acts and the regulations made under them, called 'existing statutory provisions', became subject to enforcement under it. This was important to the construction industry because the four sets of construction regulations of 1961 and 1966 made under the Factories Act 1961 became enforceable by Health and Safety Executive inspectors under the HSW Act.

Under the HSW Act several new sets of regulations have been made. Most of these apply to all industries but there are some that apply only to construction.

The effect of the UK joining the European Union was that a number of new sets of regulations were made in 1992, 1994 and 1996 to comply with various EC directives. These regulations considerably strengthen the earlier legal requirements applicable to the management and control of safety and health in construction work. The regulations brought in to implement these EC directives included the

- Management of Health and Safety at Work Regulations 1992
- Manual Handling Operations Regulations 1992
- Personal Protective Equipment at Work Regulations 1992
- Provision and Use of Work Equipment Regulations 1992
- Display Screen Equipment Regulations 1992
- Workplace (Health, Safety and Welfare) Regulations 1992

These six sets of regulations of 1992 are colloquially known as the 'six pack'. The last of these is specifically excluded from construction activities.

The EC Temporary or Mobile Construction Sites Directive has been implemented in the UK by two sets of regulations, one set of which are the Construction (Design and Management) Regulations 1994. In making the second of these the opportunity was taken to consolidate three of the four sets of Construction Regulations of 1961 and 1966 into a single set of regulations. Thus the Construction (General Provisions) Regulations 1961, the Construction (Working Places) Regulations 1966, and the Construction (Health and Welfare) Regulations 1966 have become the Construction (Health, Safety and Welfare) Regulations 1996.

It is intended that the Construction (Lifting Operations) Regulations 1961 will be consolidated within new all-industry lifting legislation in due course. Until then they will continue to apply to all construction projects.

Current construction health and safety legislation

While all regulations made under the HSW Act apply to construction work unless they are specifically excluded (e.g. the Workplace (Health, Safety and Welfare) Regulations 1992) there is a group of regulations that only apply to construction and these are described later in this chapter after a summary of the HSW Act itself.

The Health and Safety at Work etc. Act 1974

The Health and Safety at Work etc. Act 1974 (the HSW Act) covers all people at work, whether employers, employees or self-employed, except for domestic servants in private households and some public transport workers covered by other Acts. It also protects the general public where their health and safety may be affected by the work activities of others, for example, by contractors.

The HSW Act applies to England and Wales and, except for most of part III, to Scotland. Although the HSW Act does not apply to Northern Ireland, the duties imposed on employers and employees by the Health and Safety at Work (Northern Ireland) Order 1978 are substantially the same as for the HSW Act. The HSW Act is in four parts.

HSW Act part I. The purpose of part I of the Act is to

- secure the health, safety and welfare of people at work
- protect people other than those at work against risks to their health or safety arising out of the activities of people at work
- control the keeping and use of explosives or highly flammable or otherwise dangerous substances and generally prevent people from acquiring and using such substances unlawfully
- control the release into the atmosphere of noxious or offensive substances from premises prescribed by regulations.

The general duties of employers and employees are covered in sections 2–9 of part I and are summarised in the following paragraphs.

Under section 2 every employer has a duty to ensure the health, safety and welfare at work of all his employees, with specific reference to

- the provision and maintenance of plant and systems of work that are safe and without risk to health
- making arrangements for ensuring safety and absence of risk to health in connection with the use, handling, storage and transport of articles and substances
- the provision of the necessary information, instruction, training and supervision
- the maintenance of workplaces under his control in a safe condition and without risk to health; also the provision and maintenance of

means of access to and egress from places of work that are safe and without risk to health

- the provision and maintenance of working environments which are safe, without risk to health and adequate as regards facilities and arrangements for their welfare.

Also, under section 2 all employers with five or more employees must prepare a written statement of their general policy with respect to the health and safety at work of their employees. The organisation and arrangements for carrying out the policy must be included and it must be revised as often as necessary to keep it up to date. The policy statement has to be brought to the attention of all employees.

Furthermore, recognised trade unions may appoint safety representatives from among the employees, and where they have been appointed, employers must consult them with a view to making and maintaining arrangements for ensuring the health and safety at work of all employees. If requested in writing to do so, the employer must set up a safety committee whose function is to keep under review the measures taken to ensure the health and safety of employees while at work.

Section 3 requires every employer to conduct his undertaking in such a way as to ensure that people not in his employment are not exposed to risks to their health and safety. Similarly, every self-employed person must also conduct his undertaking so that he and other people (not being his employees) are not exposed to risks to their health and safety. Thus everyone, including members of the public who might be affected by the work activities, is protected under this section.

Section 4 imposes duties on those who to any extent have control over premises (other than domestic premises), the means of access, or of any plant or substances in the premises, which are used by people who are not their employees as a place of work. They must take measures to ensure that they are safe and without risk to health.

Section 5 requires those who control premises to use the best practical means for preventing the emission of noxious or offensive substances into the atmosphere and for rendering harmless and inoffensive any that do escape into the atmosphere.

Section 6 places duties on persons who design, manufacture, import or supply (by sale or hire) articles for use at work. They must

- ensure that the article is designed and constructed in such a way that it is safe and without risk to health when properly used
- make arrangements for any testing and examination that may be necessary; designers and manufacturers are also required to carry out or arrange for the carrying out of any necessary research with a view to the discovery, elimination or minimisation of any risk to health and safety to which the design or articles may give rise
- take such steps as are necessary to ensure that adequate information

will be available about the use for which the article has been designed and tested

- ensure, if they erect or install the article in any premises where it is to be used by persons at work, that it is installed or erected so as to be safe and not a risk to health when properly used.

An article is not regarded as properly used where it is used without regard to any relevant information or advice relating to its use which has been made available by a person by whom it was designed, manufactured, imported or supplied. Similar requirements apply to the supply etc. of substances for use at work.

Section 7 places duties on employees while at work. They must

- take reasonable care for the health and safety of themselves and of others who may be affected by their actions or omissions at work, and
- co-operate with their employers or anyone else as necessary to enable them to comply with their statutory duties.

Section 8 places a duty on all people, both those at work and members of the public, including children, not to interfere with or to misuse intentionally anything that has been provided in the interests of health, safety and welfare.

Section 9 prohibits an employer from charging his employees for anything done or provided for health and safety purposes to comply with a specific statutory requirement.

The remaining 45 sections of part I of the HSW Act are concerned with the powers and duties of the Health and Safety Commission and Executive, enforcement, offences and miscellaneous matters (*see also* chapter 3). However, special mention is made here of the issue of approved codes of practice.

Approved codes of practice. Under sections 16 and 17 of the HSW Act the Health and Safety Commission may approve and issue codes of practice, whether prepared by it or by some other body, for example, the British Standards Institution for the purpose of providing practical guidance on the requirements of section 2–7 of the HSW Act or of any health and safety regulations or statutory provisions.

Failure to observe the guidance provided by an approved code of practice is not in itself unlawful. However, if someone is prosecuted for a breach of any requirement of sections 2–7 of the HSW Act or related legislation and an approved code of practice is relevant to that breach, the code is admissible as evidence. Further, it is then up to the defendant to show either that the code was complied with or that what was done was as good or better than the provisions in the code.

The Health and Safety Executive usually publish an approved code of practice or guidance to each new set of regulations. Guidance is offered by the Health and Safety Executive after consultation with industry but should not be regarded as an authoritative interpretation of the law. A complete list of approved codes of practice and guidance on regulations may be obtained from HSE Publications (*see* Appendix 4).

HSW Act part II. Part II of the HSW Act describes the functions of the Employment Medical Advisory Service and its responsibility for, among other things, giving employed people and those seeking employment, information and advice on health at work.

HSW Act part III. Part III which was concerned with the making of Building Regulations has since been deleted from the HSW Act and included in the Building Act 1984. The Building Regulations are the responsibility of the Department of the Environment and the Scottish Office and they are enforced by the local authorities.

The Building Regulations are for securing the health, safety, welfare and convenience of those in or about buildings. They are principally concerned with the design and construction of buildings and the provision of services, fittings and equipment in them. They are therefore of primary relevance to building owners and designers and they are not concerned with the safety, health and welfare of those engaged in the construction, maintenance and demolition of buildings. The subjects they cover are listed in Appendix 2.

HSW Act part IV. Part IV contains various miscellaneous and general provisions, for example, amendments to the Radiological Protection Act 1970, amendments to the Fire Precautions Act 1971 and powers to repeal or modify Acts and Instruments and the extent and applications of the HSW Act.

Interpretation of the term 'reasonably practicable'. Many of the duties in the HSW Act are qualified by the term 'so far as is reasonably practicable'. For example, section 2(1) states that 'it shall be the duty of every employer to ensure so far as is reasonably practicable the health, safety and welfare at work of all his employees'. This term is not defined in the Act, but the courts have interpreted it as requiring an assessment to be made of the degree of risk which has to be weighed against the cost, in money, time and trouble of averting the risk. If the cost is high compared with the likelihood and severity of an injury (or ill health) being sustained it may well be considered not reasonably practicable to reduce the risk. However, if the likelihood of a severe injury (or ill health) being sustained is high when a certain method of work is employed then it may be considered reasonably practicable to require considerable expense to be incurred to avert the likelihood of such injury, i.e. to remove the hazard.

Whenever the expression is used without the qualification of 'reasonable', i.e. 'so far as is practicable', it is a much stricter standard and, if it is possible to obviate the hazard, then it must be done at whatever cost.

Where proceedings are instituted under the HSW Act alleging failure to comply with a requirement of duty to do something 'so far as is practicable' or 'so far as is reasonably practicable' or 'to use the best practicable means', by virtue of section 40 of the HSW Act, it is for the accused to prove that it was not practicable or reasonably practicable (as the case may be) to do more than was done or that there was no better practicable means than was used to satisfy the requirements.

Regulations that only apply to construction

Regulations made under the HSW Act or made previously and enforced under it as existing statutory provisions which only apply to construction are the

- Construction (Lifting Operations) Regulations 1961
- Construction (Head Protection) Regulations 1989
- Construction (Design and Management) Regulations 1994
- Construction (Health, Safety and Welfare) Regulations 1996

Three regulations from the Construction (General Provisions) Regulations 1961 also remain law until 31 December 1996.

Construction (Lifting Operations) Regulations 1961

There are plans to update and consolidate these regulations within new all-industry lifting legislation in due course. They deal with the construction, erection, inspection, examination and use of lifting appliances and lifting gear (tackle) on construction sites. They cover such plant and equipment as cranes, hoists, winches, piling frames, shear legs, excavators, draglines, pulley blocks, overhead runways, cableways, slings, shackles, eyebolts, hooks, wire and fibre ropes.

The regulations require that the plant and equipment be kept in good order and safe for use, and that it is used safely. They cover brakes, controls, safety devices, stability and anchorage arrangements of cranes and other lifting appliances. The testing and examination of cranes and winches etc. are covered in detail, as are the requirements for marking safe working loads, the provision of automatic safe load indicators and radius load indicators for cranes. The construction, examination, testing and safe working load for chains, ropes and lifting gear are also dealt with. The regulations also cover hoists and the carriage of people by lifting appliances.

Since the revokation in 1996 of the Construction (Working Places) Regulations 1966 the requirements concerned with non-power operated suspended scaffolds and with non-power operated boatswains chairs, cages, skips etc. have been included in the amended Construction (Lifting Operations) Regulations 1961 as Regulations 48A and 48B.

The Construction (General Provisions) Regulations 1961

These regulations are now revoked with the exception of three substantive provisions that remain in force until 31 December 1996. Regulations 42 and 43 relating to the fencing of machinery and Regulation 57 relating to the prohibited sale or hire of machinery have already been partly superseded by the Provision and Use of Work Equipment Regulations 1992 (PUWER) which currently apply to machinery first supplied after 1 January 1993. PUWER will apply to *all* machinery from 1 January 1997 and Regulations 42, 43 and 57 of the Construction (General Provisions) Regulations 1961 will then be revoked.

The Construction (Head Protection) Regulations 1989

These regulations place a duty on employers to provide suitable head protection for their employees, to maintain it and to replace it whenever necessary. They also impose a duty on every employee provided with head protection to wear it whenever required to do so.

Within the bounds of 'reasonable practicability', employers and those in control on the site are responsible for ensuring that their employees and other persons over whom they have control, wear the head protection unless there is no forseeable risk of head injury other than that likely to be caused by a person falling.

There will be very few work areas where the wearing of head protection is not required. Examples quoted in the guidance on the regulations by the Health and Safety Executive are

(*a*) in nearly completed buildings where there is no risk of objects or materials falling
(*b*) in site offices and canteens
(*c*) in the cabs of vehicles and cranes
(*d*) when all work is at ground level such as road resurfacing work and kerb laying.

Everywhere else on the site should be designated a 'head protection area' where everyone entering must comply with the regulations and any rules made in respect of them.

The regulations are also applicable to self-employed workers on site. They have a duty to provide themselves with suitable head protection and to wear it wherever required, to maintain it and to replace it whenever necessary. However, the regulations do not apply to Sikhs working on construction sites provided they wear turbans.

The Construction (Design and Management) Regulations 1994 (CDM Regulations)

These regulations gave effect in Great Britain to the 1992 EC Temporary or Mobile Worksites Directive relating to the minimum safety and health requirements at temporary or mobile construction sites. They strengthen considerably the legal requirements relating to the management and control of health and safety in construction. Duties are imposed on clients and upon designers. A planning supervisor has to be appointed by the client at the earliest practicable stage in the project and the client has also to appoint a principal contractor, who are both responsible for different aspects of health and safety management and co-ordination.

Attention is drawn to *Managing construction for health and safety* (*see* Appendix 4), *Opportunities and impositions* by Donald Bishop[1], *The CDM Regulations explained* by Raymond Joyce[2] and the video *A safer bet – an introduction to the principles of the CDM regulations* (*see* p. 241).

A health and safety plan has to be drawn up in scope before any arrangements are made for a contractor to carry out or manage any construction work. Also, a health and safety file has to be prepared containing information needed to alert contractors tendering for future maintenance, repairs, renovation and demolition to hazards that would not be obvious to a competent contractor. This file has to be given to the client at the end of the construction phase of the project and kept for future reference by anyone proposing further work on the particular structure.

A key duty of the planning supervisor is to ensure that a pre-contract health and safety plan is prepared before arrangements are made for construction. A pre-contract health and safety plan must be sufficiently developed to be included in tender documents and contain

- a general description of the work
- details of timing
- details of risks to workers
- information to enable any contractor to show he has the competence and resources required to undertake the project
- information for the principal contractor to develop the health and safety plan
- information for any contractor to understand the legal requirements placed upon him in relation to welfare arrangements.

Following the tender stage, the appointed principal contractor then becomes responsible for the development of the health and safety plan. This development should include

- arrangements for ensuring the health and safety of workers and anyone else affected by the work being carried out
- arrangements for monitoring and ensuring compliance with health and safety law
- information about welfare arrangements.

The planning supervisor has to ensure that a health and safety file is prepared, is responsible for it during the project at the end of which it must be passed to the client. However, the principal contractor and all other contractors have a duty to provide all the necessary information to enable the planning supervisor to amend and update the file. The file should include

- record drawings (as built)
- construction methods and materials
- equipment and maintenance facilities
- maintenance procedures
- maintenance manuals
- location of services and of fire-fighting systems etc.

The CDM Regulations apply to any construction project that will last longer than 30 days or will involve more than 500 person-days of construction work. They do not apply where the largest number of workers on a project at any one time is less than 5 except regarding the demolition or dismantling of a structure (as defined in the regulations) when the regulations apply irrespective of the length of time or the number of workers and similarly they always apply to design work.

To calculate the time of construction, a working day is any day on which any construction work is carried out no matter how short a duration it is or if the day is a holiday or over a weekend. A person-day is defined as one individual including supervisors and specialists carrying out construction work for one normal working shift. Therefore it should be assumed that engineers, resident engineers, architects, clerks of works and safety advisers and even those visiting are counted as 'persons' in the definition of a person-day.

Except for the requirements on designers, the regulations do *not* apply to construction work for which the local authority is the enforcing authority. This work is usually of a minor nature in such premises as shops and offices. Also, other than the requirements placed on designers, and in respect of the need to notify the Health and Safety Executive about the project, the regulations do *not* apply to work for domestic householders carried out on their own residences. But they *do* apply to commercial developers who sell domestic premises before the project is complete and who then arrange for the construction work to be carried out (e.g. a new housing estate). In such cases the developer must comply with the duties required of the client in the regulations.

The definition of 'construction work' in the CDM Regulations is as follows:

'construction work' means the carrying out of any building, civil engineering or engineering construction work and includes any of the following

(a) the construction, alteration, conversion, fitting out, commissioning, renovation, repair, upkeep, redecoration or other maintenance (including cleaning which involves the use of water or an abrasive at high pressure or the use of substances classified as corrosive or toxic for the purposes of regulation 5 of the Carriage of Dangerous Goods by Road and Rail (Classification

Packaging & Labelling) Regulations 1994, de-commissioning, demolition or dismantling of a structure,

(b) the preparation for an intended structure, including site clearance, exploration, investigation (but not site survey) and excavation, and laying or installing the foundations of the structure,

(c) the assembly of prefabricated elements to form a structure or the disassembly of prefabricated elements which, immediately before such disassembly, formed a structure,

(d) the removal of a structure or part of a structure or of any product or waste resulting from demolition or dismantling of a structure or from disassembly of prefabricated elements which, immediately before such disassembly, formed a structure, and

(e) the installation, commissioning, maintenance, repair or removal of mechanical, electrical, gas, compressed air, hydraulic, telecommunications, computer or similar services which are normally fixed within or to a structure,

but does not include the exploration for or extraction of mineral resources or activities preparatory thereto carried out at a place where such exploration or extraction is carried out.

The definition of a 'structure' has a wider application than that to which it is commonly given

'structure' means

(a) any building, steel or reinforced concrete structure (not being a building), railway line or siding, tramway line, dock, harbour, inland navigation, tunnel, shaft, bridge, viaduct, waterworks, reservoir, pipe or pipe-line (whatever, in either case, it contains or is intended to contain), cable, aqueduct, sewer, sewage works, gasholder, road, airfield, sea defence works, river works, drainage works, earthworks, lagoon, dam, wall, caisson, mast, tower, pylon, underground tank, earth retaining structure, or structure designed to preserve or alter any natural feature, and any other structure similar to the foregoing, or

(b) any formwork, falsework, scaffold or other structure designed or used to provide support or means of access during construction work, or

(c) any fixed plant in respect of work which is installation, commissioning, de-commissioning or dismantling and where any such work involves a risk of a person falling more than 2 metres.

Five key parties (firms or individuals) have specified duties under the CDM Regulations:

The client. For most construction work the term 'client' will be obvious but the regulations apply also to agents of clients and to developers. The client's duties are to

- appoint a planning supervisor and a principal contractor
- be satisfied that the planning supervisor and principal contractor are

competent. This applies also to the appointment of designers and contractors
- be satisfied that the planning supervisor, principal contractor, designers and contractors will allocate adequate resources to enable them to perform their functions under the regulations
- provide the planning supervisor with information relevant to health and safety on the project
- ensure work does not start until the principal contractor has prepared a satisfactory health and safety plan
- make the health and safety file available for inspection after completion of the project in respect of any future works in connection with the structure.

The planning supervisor. The planning supervisor has to co-ordinate the health and safety aspects of the planning and design of the project. The duties are as follows.
- To notify specified particulars of the project to Health and Safety Executive in writing preferably on form F10(Rev) if it is likely to exceed 30 days or involve more than 500 person-days. Note that until 1995 this notification was the responsibility of the contractor.
- To ensure that designers comply with their duties and that designers, if there are more than one, co-operate with each other
- To ensure a health and safety plan is prepared
- To give advice to the client on such matters as competence and resources of designers and contractors.
- To keep the health and safety file and give it to the client at the end of the project.

The designer. Designers include not only those who calculate for and prepare detailed drawings but also those who specify and who prepare bills of quantities. The designer's duty is to minimise risk to health and safety of those who are going to construct, maintain, clean, repair, dismantle or demolish the structure and anyone else who might be affected by the work. Designers must ensure that the client is aware of the client's duties under the regulations and consider the hazards and risks that might arise during construction, maintenance, cleaning, repair, dismantling or demolition.

Designers, including designers of temporary works, must avoid or reduce those risks as far as is reasonably practicable by
- considering methods that will protect workers if risks cannot be satisfactorily reduced
- ensuring that the design and drawings contain adequate information about any aspect of the project or structure or materials which might affect the health or safety of any person carrying out the construction work or cleaning work in or on the structure at any time or of any other person who might be affected by the work

- passing this information to the planning supervisor for inclusion in the health and safety plan
- co-operating with the planning supervisor and any other designers.

The principal contractor. The principal contractor not only has to take health and safety into account in contractual activities, but has to develop the health and safety plan and co-ordinate the activities of all other contractors to ensure that they comply with health and safety legislation. The principal contractor's duties are to

- develop and implement the health and safety plan
- arrange for competent and adequately resourced contractors to carry out any subcontracted work
- ensure co-ordination and co-operation between all contractors involved in the project including for example any utility company
- obtain contractor's risk assessments, ensure the seriousness of the risks have been properly evaluated and ensure that the methods to deal with the risks are incorporated in the health and safety plan
- ensure contractors have information about risks on site
- ensure workers have been adequately trained
- ensure contractors and workers comply with site rules set out in the health and safety plan
- monitor health and safety performance
- ensure workers are properly informed and consulted
- make sure only authorised people are allowed on site
- display the notification of the project to the Health and Safety Executive
- pass information to the planning supervisor for the health and safety file.

Contractors and the self-employed. The regulations require all contractors and the self-employed when they act as contractors to

- co-operate with the principal contractor and provide any information including risk assessments that might affect the health and safety of workers or anyone else
- comply with directions of the principal contractor and with the health and safety plan
- provide the principal contractor with information about any death, injury, health condition or dangerous occurrence which the contractor is required to notify or report under the Reporting of Injuries, Diseases and Dangerous Occurrences Regulations 1995 (RIDDOR 95)
- provide the principal contractor with relevant information for the health and safety file
- provide information for the health and safety file about any injuries etc., on site i.e., copies of reports required under RIDDOR 95
- provide information to their employees.

Note that the principal contractor has to check that employees have been provided with adequate information and training.

The Construction (Health, Safety and Welfare) Regulations 1996 (CHSW Regulations)

These regulations comprise the updating and consolidation into one document of three of the four sets of construction regulations of 1961 and 1966, namely the

- Construction (General Provisions) Regulations 1961
- Construction (Working Places) Regulations 1966
- Construction (Health and Welfare) Regulations 1966.

At the same time a number of new provisions were introduced to implement the requirements of Annex IV of the European Construction Sites Directive that had not been covered elsewhere.

The CHSW Regulations impose duties on employers, the self-employed (in so far as they relate to matters within their control), and employees in a similar manner to the HSW Act. Unlike the CDM Regulations, contractors and clients do not attract any duties unless they are also employers using or controlling construction workers.

The expression 'so far as is reasonably practicable' (*see* p. 18) appears often in the CHSW Regulations but it is not repeated in this summary.

The structure of the regulations is such that the various requirements are set down in broad outline but supported in detail by a number of Schedules. For example, all working places on construction sites must be made and kept safe and have safe access to and egress from them. There shall be adequate working space at each place of work. Steps must be taken to see that no person gains access to any unsafe place. Steps must also be taken to prevent anyone falling. Where any person is liable to fall a distance of 2 m or more, suitable guard rails and toe boards, barriers or other protection must be provided. This regulation is supported by Schedules 1 and 2 which provide detailed requirements for guard rails etc. and for working platforms, including their condition, stability and size and the stability of their supporting structures (scaffolding). Other means of preventing falls are supported by Schedules 3, 4 and 5, which cover personal suspension equipment, fall-arrest devices and ladders.

Those familiar with the Construction (General Provisions) Regulations 1961 and the Construction (Working Places) Regulations 1966, which were revoked upon the introduction of the CHSW Regulations, will note few changes to the older requirements except that it is now necessary to reduce the unprotected gap between the top of the toe board and the lowest guard rail and between any two adjacent guard rails from 765 mm to 470 mm or less. Also, working platforms (scaffolds) and any personal suspension equipment now need to be inspected *before* being taken into use as well as any ongoing inspection.

The regulations also spell out in detail duties in relation to

- fragile material
- falling objects
- stability of structures
- demolition or dismantling
- explosives
- excavations
- cofferdams and caissons
- prevention of drowning
- vehicles
- plant and equipment

all of which were included in the revoked regulations of 1961 and 1966 although there has been some modification and updating.

However, there are a number of requirements that became law under the CHSW Regulations with which many will not be familiar. Some of these are described below.

Traffic routes. Every construction site must be organised so that pedestrians and vehicles can move about safely (*see* p. 118) Traffic routes should be suitable for the persons or vehicles using them and suitable so as not to cause danger to those near them. Vehicles should be separated from pedestrians, or where this is not practicable means for protecting them must be provided. Where anyone is liable to be crushed or trapped effective arrangements must be made to warn them. Doors or gates leading onto vehicle traffic routes must be positioned so that pedestrians can see approaching vehicles. Any gate intended for vehicles, if unsafe for pedestrians should have an additional pedestrian gate clearly marked. Any loading bay must have a pedestrian exit point. Steps must be taken to warn drivers of low clearances and overhead obstructions by warning devices, notices or otherwise and also to warn any person riding on a vehicle of approaching danger due to inadequate overhead or side clearances.

Doors and gates. Any permanent or temporary door, gate or hatch must be fitted with any necessary safety devices. A sliding door must be prevented from coming off its track, an upward opening door must be prevented from falling back and any powered door must have a readily accessible and identifiable stop device to prevent causing injury by trapping. Powered doors must have facilities for manual operation unless they fail to open safely.

Prevention of risk from fire, etc. The risk of injury due to fire or explosion, flooding or any substance liable to cause asphyxiation must be prevented.

Emergency routes and exits. A sufficient number of these are to be provided to enable people to reach safety quickly in the event of danger. They must be kept clear of obstructions and provided with emergency lighting and signed as required.

Emergency procedures. These have to be prepared and implemented where necessary and include evacuation of the site or parts of it. The arrangements are to be made familiar to those concerned and they are to be tested from time to time.

Fire detection and fire-fighting. Fire-fighting equipment, fire detectors and alarm systems suitably located, properly maintained and subject to examination and testing at regular intervals are required where appropriate. Every person on the site has to receive instruction on the equipment they may have to use.

Welfare facilities. Sanitary conveniences, washing facilties, drinking water, accommodation for clothing, facilities for changing clothing, for rest and for eating must comply with the requirements set out in Schedule 6 of the CHSW Regulations.

Fresh air. Workplaces must have a sufficient quantity of fresh or purified air. Any plant used to provide purified air must include an effective device to give visible or audible warning of any failure of the plant.

Temperature and weather protection. Steps must be taken to ensure that the temperature of any indoor place of work is reasonable having regard to the purpose for which it is used. Protective clothing, equipment or other arrangements must be provided to give protection against adverse weather for those working outdoors.

Lighting. Every workplace and access, traffic route and dangerous opening must be suitably lit, preferably by natural light. Where artificial lighting is used, emergency lighting is required where a person is exposed to danger in the event of its failure. The colour of any artificial lighting must not adversely affect the perception of any safety sign or signal.

Good order. Construction sites must be kept in good order and every place of work kept in a reasonable state of cleanliness. Also, the perimeter of the site should be adequately signposted. Timber or other material with projecting nails must not be allowed to remain on site where it may be a source of danger.

Training. Anyone who carries out a construction activity requiring training, technical knowledge or experience to carry out the work safely must either have received such training or possess the necessary skills or be under the supervision of someone who has.

Inspection. The regulations require that a number of construction places of work listed in Schedule 7 must be inspected at the times set out in the schedule by a competent person who must be satisfied that work can safely be carried out. Such inspection does not relieve employers of the duty to see that places of work on scaffolds or in excavations, cofferdams or caissons are safe for use before people under their control use those places of work.

Reports. Inspection reports for the foregoing are required before the end of the working period concerned giving details set out in Schedule 8 of the regulations. Distribution and filing of the inspection reports are also specified. Reports or a copy must be given to the person for whom the inspection was carried out within 24 hours of completing them. Reports are not required for working platforms from which a person is liable to fall less than 2 m nor in respect of a mobile tower scaffold unless it remains erected in the same place for at least 7 days.

Further details of these regulations are to found in *The C(H, S & W) Regulations explained* by Raymond Joyce.[3]

Regulations not specific to construction

There are a number of regulations made under the HSW Act that are applicable to all industries but are nevertheless of particular importance in construction. These have been summarised in the various chapters of this book where the subject matter most closely relates. They are the:

- Control of Substances Hazardous to Health (COSHH) Regulations 1988 (*see* p. 158)
- Noise at Work Regulations 1989 (*see* p. 165)
- Reporting Injuries Diseases and Dangerous Occurrences Regulations 1985 (*see* p. 275)
- Management of Health and Safety at Work Regulations 1992 (*see* p. 194)
- Provision and Use of Work Equipment Regulations 1992
- Personal Protective Equipment at Work Regulations 1992 (*see* p. 251)
- Manual Handling Operations Regulations 1992

Other Acts and regulations that can apply to construction are listed in Appendix 1.

Test of the HSW Act in the courts – HMS Glasgow

A significant case in the administration of health and safety legislation arose from an incident during the construction of *HMS Glasgow* at Swan Hunter Shipbuilders: *R. v. Swan Hunter Shipbuilders and another* [1982] 1 All E.R. 264. Subcontractors who had no contractual relationship with the shipbuilders were working on the ship while it was being fitted out. The shipbuilders were aware that because of the use of oxygen there was a danger of fire in confined spaces but did not distribute their rule book to subcontractors. A subcontractor's employee failed to disconnect an oxygen hose and a serious fire resulted. The shipbuilders were prosecuted and found guilty of

- failing to provide safe systems of work
- failing to provide information and training
- failing to preserve the safety of employees not their own.

Subsequent to their appeal the following was held.

- The duties imposed on an employer by the HSW Act followed the common law duty of care of a main contractor to co-ordinate operations to ensure the safety of his own and other employees. The main contractor's protection lay only in the provision of the duties insofar as it was reasonably practicable to perform them. It was for the main contractor to prove that it was not reasonably practicable.
- It followed that, if it was reasonably practicable to give his own employees information and training then it was also reasonably practicable to give it to subcontractors' employees, not only for their own safety but also for the safety of main contractors' employees and anyone else.

Cases similar to *HMS Glasgow* that occur on construction sites in the future may well be taken under the CDM Regulations but the example is included to show what can happen if management fail to inform and train *all* workers on the site whoever they are employed by.

PART 2   CIVIL LAW

Basis of civil law

Contractual disagreements and compensation matters are examples of civil law which is based on the principle that like cases should be decided alike. In applying the law to a given set of facts, the court refers to the decisions of judges in decided cases, applying the doctrine of precedent. If liability is proven, in most cases damages are paid to the injured party. Civil law is mainly made up of

- the law of contract
- tort.

A contract is based on an agreement between two or more parties. The law of contract exists to protect the agreements, to ensure that the parties to them carry out the promises they have made and to provide one party with a remedy should the other fail to do what it has undertaken. The remedy for breach of contract may be liquidated damages if the parties have agreed to a pre-estimated sum to be paid in the event of a breach. If, however, as is often the case, the contract is silent regarding any sum to be paid in the event of a breach then the damages to be awarded are left to a decision of the court. This is known as unliquidated damages.

Tort, however, has nothing to do with agreement. Tort is a wrong, a breach of duty towards persons generally, such as nuisance or negligence for which the remedy is an action for unliquidated damages.

In the context of construction safety it will be necessary to consider both contract law and tort and, as will be seen, it is likely that in many cases overlaps will occur involving both forms of civil law and possibly criminal law as well.

The law of contract

Contract is the basis of all commercial law and a contract is usually made by the acceptance by one person of an offer made by another. The agreement may be arrived at in writing, by spoken word or merely by conduct or a combination of all three.

For a legal contract to exist, a number of conditions must be satisfied, one of which is that the performance of it must be legal. A contract for carrying out construction works is possibly voidable if the performance of it requires the rules of common law or a statutory provision, for example, a construction regulation, to be contravened, and it follows that tenders must cover for complying with criminal law.

Conditions of contract

For most building and civil engineering contracts, it is necessary to define the conditions under which the contractor will be paid for the work he performs. They are usually known as the conditions of contract. Because of the similarities between various types of work, a number of standard conditions of contract are prepared by trade federations, institutions and some major authorities and commercial undertakings. The ones with which most civil engineers will be familiar are the ICE Conditions of Contract. In the context of safety on site, other conditions of contract may be more or less explicit but, whatever is stated in any contract document, the provisions of criminal law cannot be circumvented or delegated.

ICE Conditions of Contract

The conditions of contract for use in connection with works of civil engineering construction, commonly known as the ICE Conditions of Contract, are approved by and may be obtained from the Institution of Civil Engineers, the Association of Consulting Engineers or the Federation of Civil Engineering Contractors. The Conditions are commonly employed in the civil engineering sector of the construction industry in the United Kingdom and, in a modified form, in many overseas territories. The two parties to the ICE Conditions of Contract are the Employer and the Contractor. An Engineer is appointed by the Employer, and his representative (the Engineer's representative or, more popularly, the resident engineer) watches and supervises the works. The Engineer may be an independent consulting engineer or a member of staff of the Employer's organisation.

There are two editions of the ICE Conditions of Contract currently in use, the fifth and sixth editions. The sixth edition, published in 1991, is basically the fifth edition re-written to remove anomalies, improve clarity and to make the document more readable by the breakdown of long clauses. The responsibilities of the parties remains the same and, apart from one change in a clause subnumber, the clauses relating to site safety are identical in the two editions.

Clauses concerning safety. Clauses that relate to site safety are clause 8(3) [was clause 8(2) in fifth edition], clause 13, clause 15(2), clause 16, clause 19(1), clause 40 and clause 71. The *Contractor's* responsibility for safety is covered by clause 8(3) clause 15(2) and clause 19(1).

Clause 8(2) states

> The Contractor shall take full responsibility for the adequacy, stability and safety of all site operations and methods of construction provided that the Contractor shall not be responsible for the design or specification of the Permanent Works (except as may be expressly provided in the Contract) or of any Temporary Works designed by the Engineer.

Clause 15(2) final sentence reads

> The Contractor or such authorised agent or representative shall be responsible for the safety of all operations.

Clause 19(1) states

> The Contractor shall throughout the progress of the Works have full regard for the safety of all persons entitled to be on the Site and shall keep the Site (so far as the same is under his control) and the Works (so far as the same are not completed or occupied by the Employer) in an orderly state appropriate to the avoidance of danger to such persons and shall *inter alia* in connection with the Works provide and maintain at his own cost all lights, guards, fencing, warning signs and watching, when and where necessary or required by the Engineer or Engineer's Representative or by any competent statutory or other authority for the protection of the Works, or for the safety and convenience of the public or others.

It is therefore made quite clear that the contractor must construct the works in a safe manner as a condition of his contract with the Employer.

The situation where the *Employer* has workmen on the site or engages other contractors is covered by clause 19(2). The Employer must require them to

(a)   have full regard for the safety of all persons entitled to be on the site
(b)   keep the site in an orderly state appropriate to the avoidance of danger to such persons.

The powers given to the *Engineer* to require the Contractor to work safely are covered in four clauses clause 13 clause 16 clause 40 and clause 71.

Clause 13(1) states

> Save in so far as it is legally or physically impossible the Contractor shall construct and complete the Works . . . to the satisfaction of the Engineer. . . . The Contractor shall take instructions only from the Engineer or (. . .) from the Engineer's Representative.

Clause 13(2) states

. . . and the mode, manner and speed of construction and maintenance of the Works are to be of a kind and conducted in a manner approved by the Engineer.

Clause 16 states

. . . and the Engineer shall be at liberty to object to and require the Contractor to remove from the Works any person employed by the Contractor in or about the execution of the Works who in the opinion of the Engineer misconducts himself or is incompetent or negligent in the performance of his duties or fails to conform with any particular provisions with regard to safety which may be set out in the Specification or persists in any conduct which is prejudicial to safety or health and such persons shall not be again employed upon the Works without the permission of the Engineer.

Clause 40(1), abridged, states

The Contractor shall on the written order of the Engineer suspend the progress of the Works or any part thereof for such time and in such manner as the Engineer may consider necessary and shall during such suspension properly protect and secure the work so far as is necessary in the opinion of the Engineer. . . the Contractor shall be paid . . . under this Clause except to the extent that such suspension is:-

(c)   necessary . . . for the safety of the Works . . .

Therefore, if the Engineer is concerned about the manner of construction and safety, the offending employee(s) can be sent off the site (clause 16) or perhaps work can be suspended (clause 40). However, whether the phrase 'for the safety of the Works' in clause 40 is meant to include 'health and safety of the workers' is open to question. The subject is discussed later.

Clause 71 deals with the CDM Regulations (*see* p. 20).
Clause 71(3) states

Any action under the CDM Regulations taken by either the Planning Supervisor or the Principal Contractor and in particular any alteration to the Health and Safety Plan shall be deemed to be an Engineer's instruction pursuant to Clause 13 (*see* earlier).

Clause 71(3) continues, stating that the Contractor will not be entitled to additional payment or extension of time as a result of these instructions unless he can show that they could not reasonably have been foreseen.

The ICE Conditions and the HSW Act. A number of editions of the ICE Conditions of Contract have been published, the first in 1945 and the fifth in 1973, one year before the HSW Act. As a result there are some aspects of the fifth edition that are not entirely compatible with current safety legislation. The same applies to the sixth edition which is little different from the fifth.

While it is perfectly true that the performance of the contract must comply with statute law it is not only the Contractor who is liable to prosecution under the HSW Act if an accident should occur! (*see* Fig. 6.1). This is why some clients have added additional clauses, when using ICE Conditions as a basis for contract documents, to strengthen the role of the Engineer in terms of site safety.

However, some of these weaknesses are now covered by statute law under the CDM Regulations. Ensuring the competence of contractors and subcontractors in health and safety terms and developing the health and safety plan to ensure effective site management during the construction phase are now statutory requirements and the Principal Contractor has the duty to ensure that these matters are implemented even though the ICE Conditions make no mention of them. Clause 71 of ICE Conditions determines that instructions to the Contractor from the Principal Contractor are to be taken as though they had come from the Engineer.

Nevertheless there are two matters that compilers of contract conditions should address. One is the contractual relationship, if there is to be one, between the Engineer's Representative and the Principal Contractor. Left undefined, the administration of the Contract by the Engineers Representative could be confused.

The other matter concerns the wording of Clause 40. As it is printed, the Engineer has powers to suspend the Works or part, if it becomes necessary for the *safety of the Works*. This phrase should be expanded to read *for the safety of the Works or for the health or safety of persons on or affected by the Works*.

Contracts for professional services

Contracts to carry out the design, specification and supervision of civil engineering work range from a simple exchange of letters to legally prepared conditions of engagement signed by both parties on an offer and acceptance basis. The Association of Consulting Engineers (ACE) has drawn up standard Conditions of Engagement to form the basis of the agreement between the client and the consulting engineer for five different types of employment, including report and advisory work, civil, mechanical and electrical and structural engineering work, and engineering services. Section 5 of these conditions is concerned with care and diligence

> 5.1 The Consulting Engineer shall exercise all reasonable skill, care and diligence in the discharge of the services agreed to be performed by him etc.

Even if the ACE Conditions of Engagement are not used and the terms of appointment do not specifically mention that the engineer should use proper skill and care, the law will imply that he is bound to do so.

The question of professional skill and care has been considered by the High Court in *Greaves and Co. (Contractors) Ltd* v. *Baynham Meikle and Partners* (1975) 1 W.L.R. 1095. In this case, professional design services had been provided to the contractor to assist in a design-and-build project. The first floor failed because of insufficient attention to vibration effects of live loading on the structure and the consulting engineers were held to be liable for breach of contract. This decision is important to all professional engineers since it confirms that, if there are special factors that are known or ought to be known to a professional man and he is warned, perhaps in a code of practice, about dangers associated with them, then they constitute part of the ordinary skill expected of a competent engineer. In the special case of design-and-build contracts there is an obligation not only to design a structure that is safe, but also it must be reasonably fit for the purpose for which it is required unless the designer's obligation is expressly limited to exercising reasonable skill, care and diligence.

Under the CDM Regulations, statutory duties are imposed upon clients and their planning supervisors, designers and principal contractors. Clients have to ensure, possibly with the advice of their planning supervisors, that from the standpoint of health and safety the designers and principal contractors they engage are competent and that they allocate adequate resources to the project. Designers, among other things have to give advice to their clients on the client's duties under the CDM Regulations (*see* p. 20).

Any injury or dangerous occurrence stemming, for example, from an error at the planning stage of the project may well involve the client or his appointed agent, his planning supervisor or a designer or indeed all three in a prosecution by the Health and Safety Executive. It is advisable that the statutory duties of both parties to a contract for professional services are fully recognised in the contract document lest there be any misunderstanding that eventually leads to an accident. However the contract is worded, neither party can assign its criminal law duties e.g. duties under the CDM Regulations, to another person.

There is a good case for clients to place separate contracts with their planning supervisors and their designers even though they may be employed by the same firm of professional advisers. Should the occasion arise, clients would then be able to demonstrate that they discharged their duty to appoint only the competent and adequately resourced professional to each activity.

Limitation

The period for which a contractor or consulting engineer remains liable under the law of contract is six years in the case of an action on a simple contract and twelve years for a contract under seal. The time runs from the date when the cause of action accrues. In the case of contract the cause of action accrues at the moment of the breach of contract and it is irrelevant that damage is not suffered until a later date.

The limitations to the period of contract do not apply in the case of a fraudulent action. In the case of *Clark* v. *Woor* (1965) 1 W.L.R. 650, the defendant, a builder, contracted to build a bungalow with special bricks. In fact he used bricks that were inferior to those specified and eight years later in 1961 the plaintiff noticed some flaking of the bricks and discovered the breach of contract. The judge found that the plaintiffs relied on the builder, who was experienced, for honest performance of his contract. The builder had committed a fraud and consequently the plaintiff's action was not barred by the eight-year interval. The damages were calculated at prices in 1961, not in 1965 when the action was heard.

Failures involving injury at work

In the case of a site accident or failure that causes injury it is more than likely that any action for civil damages brought by the injured party or his dependants against a contractor or designer would be brought under the tort of negligence.

Tort

The rule of common law is that everyone owes a duty to everyone else to take reasonable care not to cause them foreseeable injury. Hence a contractor will be liable for failing to take reasonable care to protect those on site from foreseeable injury, and a consulting engineer or architect will be liable for failing to take reasonable care to prevent injury by reason of a faulty design. The basis of this liability is the tort of negligence, and the remedy or redress is an assessment by the courts of the damages suffered.

Negligence

Negligence within the context of health and safety is an independent tort consisting of a breach of a legal duty to take care resulting in damage. In the famous case of *Donoghue* v. *Stevenson* (1932) A.C. 562, Lord Atkin said, 'You must take reasonable care to avoid acts or omissions which you can reasonably foresee would be likely to injure your neighbour. Who then, in law, is my neighbour? The answer seems to be – persons who are so closely and directly affected by my act that I ought reasonably to have them in contemplation as being so affected when I am directing my mind to the acts or omissions which are called in question'.

Limitation in tort

Whereas under contract law the period for which a contractor or engineer remains liable is six years in the case of an action on a simple contract and twelve years for a contract under seal, in tort the usual period of limitation is six years.

The cause of action in the case of tort accrues at the moment when the plaintiff would be entitled to succeed in an action against the defendant.

In the case, for example, of negligence, this moment will be as soon as the damage occurs. When damage has arisen the cause of action accrues even though the plaintiff may be unaware that the damage has in fact occurred.

In the case of personal injuries, legal proceedings must be started within three years of the accident. After that time they are statute barred. This time may be extended in special cases, for example, to those who are found to be suffering from certain diseases after the limitation of time has expired.

The period of limitation has been modified to some extent by the Latent Damage Act 1986. The main effect of this Act is that, in negligence cases involving latent defects, the existing six-year period of limitation is subject to an extension to allow the plaintiff three years from the date of discovery or reasonable discovery of significant damage. However, a long stop is applicable, which bars a plaintiff from initiating court action more than fifteen years from the defendant's breach of duty.

Negligence involving injury

The incident at Abbeystead near Lancaster when an explosion in a completed pumphouse killed 16 people and injured many others illustrates the process of the law of tort in civil engineering. Methane had entered the valve house from an underground void.

The Water Authority, consulting engineer and contractor were sued in the High Court for negligence by a consortium of 31 victims and relatives. The contractual relationships between the defendants were irrelevant as they were sued in tort and the plaintiffs claims turned on the common law duty of care.

Following an appeal by a majority of two to one, the judges found the consulting engineer solely to blame for the accident. The Appeal Court ordered the consultant to pay the costs of all parties for both the Appeal and the previous High Court rulings.

Employer and employee

The relationship between employer and employee is covered in a form of contract known as an employment contract but liability for injuries, disease or death at work is normally dealt with under tort. In order to show that an employer was negligent, an employee must establish that the employer

- owed him a duty of care
- acted in breach of that duty
- caused injury, disease or death to the employee.

The duties of employers at common law were identified by the House of Lords in the leading case of *Wilson's and Clyde Coal Co.* v. *English* (1938) A.C. 57. All employers must provide a competent workforce, adequate materials and a proper system and effective supervision. The employer's duty is not absolute; he does not warrant that his premises and the conditions of work are safe. He is bound only to take all reasonable care, but if he delegates the performance of this duty to an independent contractor the employer still remains liable for the negligence of that contractor.

The common law duties of employers are owed only to employees. They are not owed to employees of independent contractors (as distinct from subcontractors) who may be working on the same site, even though breach of duty towards persons other than employees gives rise to criminal liability under the HSW Act. The fact that a person styles himself as self-employed does not mean that he cannot be an employee, and it follows that a contractor may be held liable for injuries suffered by a labour-only subcontractor in spite of the fact that the latter regarded himself as self-employed.

Disobeying safety orders

An employee who disobeys safety instructions can be dismissed after due warnings, or instantly in the case of gross misconduct. To ensure that dismissals on these grounds are reasonable, it is prudent for every employer to include in employees' terms of engagement, or in the health and safety policy statement, a warning to the effect that disobeying safety instructions and failing to co-operate with their employer in fulfilling their statutory duty on health and safety are subject to reprimand and, in repeated or extreme cases, dismissal.

In the case of employment injuries, the defence by employers of *volenti non fit injuria* (to one who is willing no harm be done) is rarely successful, the courts taking the view that the dangers of employment are not voluntarily accepted by workers; rather that employment is an economic necessity. However, an employer may be successful in reducing damages on the grounds of contributory negligence by the employee.

Vicarious liability

Vicarious liability is the principle under which a person is held liable for the tort of another committed against a third party. Under this doctrine an employer is liable for the torts of his employees committed by them in the course of their employment whether authorised by their employer or not. If an employee, while acting in the course of employment, negligently injures another employee or an employee of another company working on the site, or even a member of the public, the employer will be liable for the injury. Vicarious liability is the ground on which most claims for injury-causing accidents are successful.

3 Enforcement of health and safety law

Application to Great Britain

Before 1 January 1975 when the Health and Safety Commission and Executive were established, HM Factory Inspectorate within the then Department of Employment was responsible for implementing safety, health and welfare law applicable to the construction industry. Although since that date the Factory Inspectorate still has that responsibility, it now operates within the framework of the Health and Safety Executive. There are, in addition, some aspects of safety in civil engineering (e.g. railways, offshore and nuclear installations) which are covered by other inspectorates, and others (e.g. radiological protection) that are administered by the Commission on behalf of other ministries.

Health and Safety Commission

The HSW Act makes provision for the establishment of a Health and Safety Commission. This is the authority in England, Scotland, Wales and offshore that is responsible to the Secretary of State for Employment for all health and safety at work matters. Its duties include promoting the objectives of the HSW Act, carrying out and encouraging research and training, providing an information and advisory service and putting forward to government proposals for regulations under the Act. The Health and Safety Agency for Northern Ireland has similar duties and powers in Northern Ireland.

The Act requires the Commission to consist of a full-time Chairman appointed by the Secretary of State for Employment, and between six and nine part-time members. Three of the part-time members are required to represent employers' organisations, three employees' organisations and three local authorities and other appropriate organisations.

Health and Safety Executive

The HSW Act also makes provision for the establishment of the Health and Safety Executive. This body consists of three full-time members, one of whom is the Director General. They are appointed by the Health and Safety Commission with the approval of the Secretary of State.

One of their duties is to make adequate arrangements for the enforcement of the relevant statutory provisions. Thus the enforcement of health and safety at work law is undertaken by the staff appointed by the Health and Safety Executive

The other bodies which assist the Commission in a similar way include the local authorities, which have a special responsibility for safety and health in offices, shops and warehouses.

Figures 3.1 and 3.2 show the reporting relationships of the Health and Safety Commission, the Executive and its various Inspectorates, and regional organisation. Appendix 3 gives the Health and Safety Executive addresses.

HM Factory Inspectorate. The Factory Inspectorate of the Health and Safety Executive is the enforcement authority responsible for the statutory requirements for safety, health and welfare on construction sites. It also has other enforcement duties, for example for factory premises, hospitals, schools, universities and fairgrounds. It is supported and assisted in this by all the other divisions of the Health and Safety Executive, but particularly

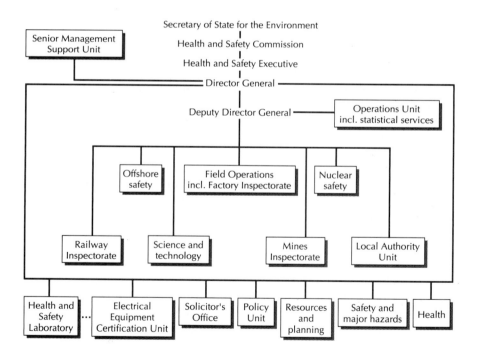

Fig. 3.1. Reporting relationships within the Health and Safety Executive

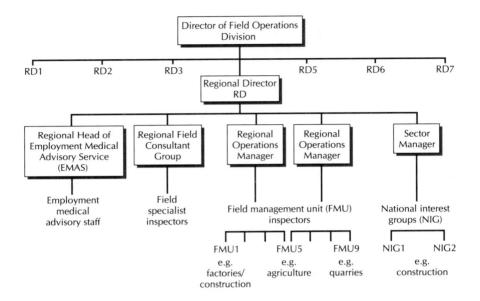

Fig. 3.2. Regional structure of the Health and Safety Executive

by the Science and Technology Division, the Health and Safety Laboratory and the Health Division.

For field inspection purposes of the Factory Inspectorate, Britain is subdivided into twenty geographical areas, each under the control of a regional operations manager (*see* Fig. 3.2 and p. 43). Regions are staffed by inspectors who are involved in the day-to-day inspection and advisory activities. In each area one group of inspectors, under the direction of a senior inspector, is solely concerned with the activities of the construction industry. They carry out unannounced inspections of sites and investigations of accidents or dangerous occurrences on site, and provide advice on compliance with the appropriate law to contractors and others who have specific interests in construction activities, for example manufacturers of construction plant.

In addition, the National Interest Group for Construction is located at 1 Long Lane, London SE1 4PG. This co-ordinates the activities of the construction groups in the areas and also acts as the focal point for discussion between the construction industry and the Health and Safety Executive and provides the secretariat for CONIAC (*see* p. 50).

The Science and Technology Division. Staffed by professionally qualified engineers, architects, chemists, occupational hygienists, physicists and microbiologists, this division has professional responsibility for a field force in addition to its larger headquarters units. For this purpose Britain is divided into seven geographical areas each covered by a Field Consultant

Group (FCG)(*see* Fig. 3.2). Conventionally each FCG (except the one covering the north-east of England) is housed within one of the Factory Inspectorate regional offices and serves two or three adjacent regional operational areas. In addition to mirroring the headquarters discipline units, they manage the scientific and laboratory analysis and sampling work. They provide immediate professional and technical support to the field inspectors.

Staff in headquarters units of the Science and Technology Division specialise in various topics within their discipline, e.g. in the Construction Unit, tunnelling, earthworks, structural engineering, demolition work including the use of explosives, falsework, scaffolding and construction plant. They provide support and authoritative professional advice to the Health and Safety Executive's field and policy divisions in addition to contributing nationally and internationally in safety and health matters as members of British and international standards committees.

The Science and Technology Division includes the Explosives and Flammables Unit which is concerned with the manufacture, transport, handling and security of explosives, and the Major Hazards Assessment Unit. The unit develops techniques and methods for assessing quantitatively hazards and risks arising from major hazards. It formulates advice for use by planning authorities for the siting of new major hazard premises and for the control of developments near existing major hazard sites. Major hazard sites may briefly be described as plant or premises which contain sufficient quantities of hazardous substances with the potential to cause major accidents from the release of flammable or toxic substances (*see also* p. 230).

The Health Division. This division of the Health and Safety Executive is staffed by doctors and nurses who specialise in occupational medicine. They have both a headquarters and a field presence. In the field they are based in area offices and provide advice on health at work; they also carry out surveys and investigations on occupational ill health. The Employment Medical Advisory Service (EMAS) provides free advice at the request of any employed or self-employed worker, trade union representative, employer or medical practitioner on the effects of work on health, together with guidance on the placement and return to work of people with health problems.

Field Operations Division. Seven regional directors of field operations are each responsible for a number (between two and four) of the 20 areas each under its regional operations manager. The factory, agricultural and quarries inspectors form part of the regional operations manager's team.

Each region also includes EMAS teams and one field consultant group. Scientific support for field inspectors is provided on an outside contract basis but managed by the head of the FCG. Each regional director reports to the Director of Field Operations, who in turn reports to the deputy Director General of the Health and Safety Executive. The field management structure is shown in Fig. 3.2.

Powers of the Factory Inspectorate

The task of the inspectors within the construction groups is to ensure compliance with the HSW Act by all those engaged in construction operations within their particular Factory Inspectorate area. They have wide powers to enter sites and premises to carry out inspections, examinations and investigations. These powers include the right to

- enter premises or construction sites at any reasonable time to conduct inspections and examinations for the purpose of ensuring compliance with any of the relevant statutory provisions
- take with them any other person who has been authorised by the Executive (e.g. another inspector, a scientist or a photographer), and any equipment or materials required to carry out their work
- carry out investigations and examinations (e.g. in respect of an accident)
- direct that any premises or site (or any part of them) and/or anything in them such as plant, machinery, substances, etc. be left undisturbed until their investigations have been completed
- take measurements, photographs and any recordings they consider necessary for the purpose of any examination or investigation they undertake
- take samples of articles, substances and the atmosphere in or in the vicinity of the premises and site they are empowered to enter
- require any article or substance to be dismantled or subjected to any process or test if they consider it to have caused or be likely to cause danger to health or safety; however, this should be carried out without damaging or destroying the article or substance unless necessary for the purpose of carrying into effect any statutory provision
- take possession of any article or substance mentioned in the previous item and keep it for as long as is necessary in order to
  - examine or dismantle it or both
  - ensure it is not tampered with before their examination has been completed
  - ensure it is available for use as evidence in legal proceedings in which event the inspector is required to leave with a responsible person on the premises, a notice describing the article or substance, sufficient to identify it and stating that the inspector has taken it; if, in the case of a substance, it is possible, he shall only take a sample, marked in a manner sufficient to identify it
- seek information from any person whom they consider may be able to help with any examination or investigation they are conducting, require the person to answer questions (in respect of the investigation etc.) and require the person to sign a declaration of the truth of the answers; this may only be carried out in the presence of a person nominated by the person being questioned and any person the in-

spector may allow to be present; however, the answers given to the inspector cannot be used in evidence in any proceedings against that person for a breach of the Act or relevant regulations; also they cannot be used in evidence for similar proceedings against the husband or wife of a person
- require the production of, inspect and take copies of or copy any entry in
  - books, documents, registers, etc. that are required to be kept by the relevant statutory instruments
  - any other books, documents, drawings, etc. which it is necessary for them to see for the purposes of any examination or investigation they are undertaking
- require any person to provide such facilities and assistance with matters or things within that person's control or in relation to which that person has responsibilities as are necessary to enable the inspectors to exercise the powers conferred on them
- exercise any other powers necessary for the purposes of their work
- seize and cause to be rendered harmless, by destruction if necessary, any article or substance in the premises or on the site that they consider, in the circumstances seen, could be a cause of imminent serious personal injury.

Methods to ensure compliance

Various ways are used by the inspectors in the field to secure compliance with legal requirements. Many involve routine inspection of sites to check what is occurring against what is required by the law. The visits are made without giving any prior warning to the contractor or site manager.

Advice and persuasion. The inspector may find a contravention of, for example, a construction regulation. In some cases the contractor may not be aware of the particular requirement or how it may be complied with. The inspector will then give advice on ways in which the obligation may be met. However, the contravention may be an oversight by the contractor or something unknown to the person in charge. Again, advice and persuasion to correct the fault will be the inspector's first action.

If the inspector finds a number of matters requiring attention, again verbal advice will be given on what is wrong and how to put it right. This may then be put in writing to the firm concerned, indicating also the various breaches of the law.

Notices. The inspector is provided with a particularly quick and effective means of dealing with breaches of the law and hazards that may be found during site inspections: the issue of an official notice. There are two types: the Improvement Notice and the Prohibition Notice.

An *Improvement Notice* can only be issued where an inspector considers that there has been a contravention of a statutory provision and it is likely that the contravention will continue or be repeated, or at the time of the

Health and Safety Executive
Health and Safety at Work etc Act 1974, Sections 21, 23 and 24

Serial number

ǀ ANO/1/11.1.89

Improvement notice

Name DAWKIN EXCAVATION AND MUCKSHIFT LTD

Address 9 EARTHTON MANSIONS, WAKEFIELD

Trading as*

Inspector's full name   I,   ANTHONY NIGEL OTHER

Inspector's official
designation   one of her Majesty's Inspectors of   Factories
Being an Inspector appointed by an instrument in writing made pursuant to section 19 of the said Act
and entitled to issue this notice

Official address   of Health and Safety Executive, The Yard, Wakefield, West Yorkshire.

Telephone number 012-345-6789

hereby give you notice that I am of the opinion that

Location of premises   at your Gas Street, Wakefield, construction site.
or place of activity

you, as an employer/a̶ ̶s̶e̶l̶f̶-̶e̶m̶p̶l̶o̶y̶e̶d̶ ̶p̶e̶r̶s̶o̶n̶/̶a̶ ̶p̶e̶r̶s̶o̶n̶ ̶w̶h̶o̶l̶l̶y̶ ̶o̶r̶ ̶p̶a̶r̶t̶l̶y̶ ̶i̶n̶ ̶c̶o̶n̶t̶r̶o̶l̶ ̶o̶f̶ ̶t̶h̶e̶ ̶p̶r̶e̶m̶i̶s̶e̶s̶/̶o̶t̶h̶e̶r̶ *
are contravening/h̶a̶v̶e̶ ̶c̶o̶n̶t̶r̶a̶v̶e̶n̶e̶d̶ ̶i̶n̶ ̶c̶i̶r̶c̶u̶m̶s̶t̶a̶n̶c̶e̶s̶ ̶t̶h̶a̶t̶ ̶m̶a̶k̶e̶s̶ ̶i̶t̶ ̶l̶i̶k̶e̶l̶y̶ ̶t̶h̶a̶t̶ ̶t̶h̶e̶ ̶c̶o̶n̶t̶r̶a̶v̶e̶n̶t̶i̶o̶n̶ ̶w̶i̶l̶l̶
c̶o̶n̶t̶i̶n̶u̶e̶ ̶o̶r̶ ̶b̶e̶ ̶r̶e̶p̶e̶a̶t̶e̶d̶* the following statutory provisions;

Construction (General Provisions) Regulations 1961, Regulation 5.

The reasons for my said opinion are: Your company is a contractor undertaking
works of engineering construction to which the above Regulations
apply.   The company normally employs more than 20 persons in such
work and has not specifically appointed in writing one or more
persons experienced in such works and suitably qualified for the
purpose to be specially charged with the duties of

(a)   advising the company as to the observance of the requirements
       for the safety under the Factories Act 1961 and subordinate
       legislation and other safety requirements and

(b)   exercising a general supervision of legal requirements and
       promoting the safe conduct of the work on the company's sites.

and I hereby require you to remedy the said contraventions or, as the case may be, the matters
occasioning them by 11 March 1989   (and I direct that the measures specified in the
Schedule which forms part of this Notice shall be taken to remedy the said contraventions or matters)*

Signature A N Other   Date   11.1.1989

A̶n̶ ̶i̶m̶p̶r̶o̶v̶e̶m̶e̶n̶t̶ ̶N̶o̶t̶i̶c̶e̶ ̶i̶s̶ ̶a̶l̶s̶o̶ ̶b̶e̶i̶n̶g̶ ̶s̶e̶r̶v̶e̶d̶ ̶o̶n̶

o̶n̶

x̶ ̶c̶o̶p̶i̶e̶s̶ ̶o̶f̶ ̶t̶h̶e̶ ̶r̶e̶l̶e̶v̶a̶n̶t̶ ̶s̶c̶h̶e̶d̶u̶l̶e̶ ̶a̶r̶e̶ ̶a̶t̶t̶a̶c̶h̶e̶d̶

LP1 (rev 1/86)   *See notes overleaf*   *please delete as appropriate*

HEALTH AND SAFETY EXECUTIVE
Health and Safety at Work etc. Act 1974, Sections 21, 22, 23, and 24

Schedule

Serial No. **I/P**

If any of the following measures are put into effect then the Notice will be deemed to have been complied with.

Either:

1   Any person or persons in the company with suitable experience of site work should be sent on the "Construction Safety" Course at XYZ Polytechnic or the "Safety Supervisor" Course of the South Riding Safety Group and then appointed in writing as a company safety supervisor with responsibility to visit regularly and inspect all company construction sites

Or:

2   At outside specialist safety management company shall be appointed by Dawkin Excavation and Muckshift Ltd to undertake on Dawkin's behalf the duties of Regulation 5.

Or:

3   Other action shall be taken by the company of equivalent standard to 1 or 2 above to ensure compliance with Regulation 5.

A N Other   11.1.1989

Fig. 3.3. (above and facing page). Example of a Health and Safety Executive improvement notice

visit a contravention (or more than one) is seen. An example of an Improvement Notice is given in Fig. 3.3. The inspector has to include in the notice

- the location of the site or place where the contravention occurred
- the particular provisions contravened
- the reasons why in his opinion the provisions have been contravened
- a realistic date by which the matters are expected to be remedied.

The contractor may appeal against the notice but this must be done within 21 days of the date of issue. The appeal has to be forwarded to an Industrial Tribunal (the address of the Secretary of the Tribunals, together with the procedures for lodging an appeal, are given on the notice) which after consideration may cancel, confirm or modify the notice. When an appeal is lodged the requirements of the notice are suspended pending the results of the appeal.

A *Prohibition Notice* may be used only when the inspector considers that there is a risk of serious personal injury arising from an activity or operation being carried on or about to be carried on. It has the effect of preventing an activity or operation being carried out or continued until the requirements specified by the inspector in the notice have been carried out and any associated contraventions remedied. An example of a Prohibition Notice is given in Fig. 3.4.

The notices can be issued to have an immediate or a deferred effect. The immediate type may only be issued if the inspector considers that the risk of injury is imminent. If it is of the deferred type then the prevention of the activity or operation becomes effective from the date stipulated in the notice unless the requirements of the inspector have been complied with. With both types of notice, when the contractor has remedied the matters and any associated contraventions that are stipulated, the notice ceases to have any effect. As with Improvement Notices, an appeal may be lodged through an Industrial Tribunal, but Prohibition Notices remain in force until the appeal has been heard.

Prohibition Notices are served in the anticipation of danger and may be issued without there being a contravention of any particular regulation. However, if there has been a legal contravention then the details must be stipulated in the notice.

Both the immediate and deferred notices may, but need not, include directions for remedying the faults or contraventions which the inspector requires to be carried out.

Prosecution. The ultimate course taken by inspectors to uphold the law is to prosecute for contravention. Anyone who does not comply with any part of the HSW Act or its relevant regulations can be prosecuted in the Magistrates' Court (or in the Sheriff's Court in Scotland). The maximum fine which may be imposed for each offence if the defendant is found guilty is £20 000 and in some specified circumstances, such as disobeying Prohibition Notices, to a term of imprisonment. However, serious cases may be brought on indictment before the Crown Court. In such cases if the defendant is found guilty a prison sentence of up to two years or a fine (the amount unlimited) or both may be imposed.

Consultation. Area inspectors also act in a consultancy capacity. They visit, by invitation or otherwise, contractors' offices to discuss problems, and to advise on legal requirements and on how best to meet these obligations. They attend local safety group meetings and give talks and advice to members on health and safety matters. They also call together representatives of specialist contractors (e.g. roofers, demolition contractors and scaffolders) in their areas to discuss ways of meeting their legal obligations.

Other activities

In addition to the work of inspectors in the field, there is continuous activity by inspectors and others from the various divisions at headquarters of Health and Safety Executive to help contractors and others concerned with

Health and Safety Executive
Health and Safety at Work etc Act 1974, Sections 22, 23 and 24

Serial Number

P ANO/1/11.1.89

Prohibition notice

Name DAWKIN EXCAVATION AND MUCKSHIFT LTD

Address 9 EARTHTON MANSIONS, WAKEFIELD

Trading as*

Inspector's full name   I,   ANTHONY NIGEL OTHER

Inspector's official designation   one of Her Majesty's Inspectors of Factories
Being an Inspector appointed by an instrument in writing made pursuant to section 19 of the said Act and entitled to issue this notice

Official address   Health and Safety Executive, The Yard,
of   Wakefield, West Yorkshire

Telephone number 012-345-6789

hereby give you notice that I am of the opinion that the following activities namely:

The digging of trenches and the laying of gas pipes

Location of premises or place of activity   which are being carried on by you/ ~~by or on your behalf by you/under your control~~* at

Gas Street, Wakefield

involve ~~or will involve~~* a risk of serious personal injury, and that the matters which

give rise / ~~will give rise~~* to the said risk(s) are: Persons are working in a 4 metre deep excavation the sides of which are not supported or battered back to prevent danger of collapse

and that the said matters involve / ~~will involve~~* contravention of the following statutory provisions:

The Construction (General Provisions) Regulations 1961
Regulation 8(1)

because   An excavation deeper than 1.2m was not suitably supported or the sides adequately sloped to prevent danger to persons employed in the excavation from a fall or dislodgement of earth and rock.

and I hereby direct that the said activities shall not be carried on by you or under your control immediately/~~after~~*   unless the said contravention(s)* and matters have been remedied.

~~I further direct that the measures specified in the schedule which forms part of this notice shall be taken to remedy the said contravention(s)* or matters.~~

Signature   A.N. Other   Date 11.1.1989

* ~~A Prohibition Notice is also being served on~~

~~or~~

~~related to the matters contained in this notice.~~

LP2 rev (1/88)   *See notes overleaf*   *delete as appropriate*

Fig. 3.4.   Example of a Health and Safety Executive prohibition notice

the construction industry to reduce accidents and ill health on site. This includes

- the collection, analysis and publication of accident statistics

- the production of guidance notes and other information (*see* p. 238 and Appendix 4)
- liaison with the various national trade and industry federations and trade unions
- representation on British Standard committees, on the Safety Panel of the Institution of Civil Engineers, the International Tunnelling Association Committee and the British Tunnelling Society Committee, for example
- liaison with other professional institutions.

CONIAC. A particularly important activity of the Factories Inspectorate is its involvement with the Construction Industry Advisory Committee. The Committee consists of representatives from both the employer and the employee sides of industry, and its function is to advise the Health and Safety Commission about the protection of people in the industry and the public from hazards to health and safety arising from construction activities. It is chaired by a senior factory inspector and attended by another who is head of the National Industry Group for construction, the specialist inspector who is head of the construction unit within the Science and Technology Division, and a senior doctor from the Employment Medical Advisory Service. CONIAC has been responsible for a number of useful publications, some of which are listed in Appendix 4. The Institution of Civil Engineers is represented on CONIAC.

4 Hazards of construction and their prevention

From the analysis of injuries and causes of ill health described in chapter 1 it will be evident that providing safe working places and safe systems of work by effective safety management are the two most important contributions engineers and other professionals can make to reduce the number of deaths and injuries in the construction industry. In this chapter the common causes of accidents are examined and methods suggested for their prevention.

The chapter is in two parts, covering physical injury hazards and health hazards, which are further subdivided into sections as follows.

- Physical injury hazards
  - excavations
  - scaffolding
  - falsework
  - structural framework
  - roofwork
  - cranes
  - transport, mobile plant and roadworks
  - tunnelling
  - sewers and confined spaces
  - demolition and contaminated sites
  - work over water
- Health hazards
  - chemical
  - physical
  - biological

The particular health hazards associated with demolition are covered under the demolition and contaminated sites section of physical injury hazards.

The hazards of injury and ill health resulting from manual handling operations are dealt with as physical hazards.

PHYSICAL INJURY HAZARDS

Excavations

This section examines the hazards and protective measures concerned with three types of excavation work: trenches; basements and wide excavations; and shafts.

Trenches

The main cause of accidents in trench work is the collapse of the sides burying or partially burying those in the trench (Fig. 4.1). About one in five of this type of accident results in death. The other major causes are materials falling into the trench striking those below and also both plant and people falling into the excavation.

Some of the reasons why the collapse of a trench is so dangerous are as follows.

- In a trench the working space is very confined, and falling materials cannot easily be avoided.
- Collapse occurs quickly and usually without warning.
- The likelihood of locating, uncovering and rescuing any buried worker in time is not good. A hydraulic excavator should never be used to clear away the debris for fear of causing further serious injury to the buried person. Even hand tools such as spades have to be used with great care to avoid further injury.
- The rescuers themselves have to be safeguarded against further collapse and the provision and positioning of extra shoring for this takes valuable time.

Causes of trench collapse. The most common causes of trench collapse are

- mechanical failure of the soil
- breakdown in strength of the ground by the ingress of water
- vibration and weight of vehicles and plant
- storage of materials and equipment near the edge of the trench
- variation in the nature of the ground
- previously excavated ground
- the sides of the trench being struck by heavy loads.

When a trench is cut, lateral support on the cut planes is removed and *mechanical failure* occurs if the ground is unable to support its own weight. Similarly, slipping and collapse may occur in rock excavations when bedding planes dip steeply across the line of the excavation. The use of

Fig. 4.1. Typical trench collapse due to inadequate shoring, often a cause of fatal injuries

explosives in the excavation process may well trigger off a slipping movement.

The *ingress of water* not only removes particle adhesion or lubricates slipping planes but also increases the weight (pressure) on the unsupported walls of the trench. Water penetration into cracks because of shrinkage of clays in hot dry weather causes softening and a rapid loss of strength. Another effect of water percolation, during winter months, is that the water may freeze and cause a breakdown of strength, thus increasing the size of any cracks or fissures.

The *movement or operation of vehicles and plant* (e.g. dumpers and cranes) close to the edge of a trench produces a heavy surcharge load on the unsupported sides which they are unable to resist. Vibration due to movement of plant may cause particle movement or even displacement of stones and small boulders which in turn leads to instability. For trenches in or adjacent to roadways, normal traffic is a particular hazard.

Excavated spoil, pipes, bricks and timbering equipment *stored too close to the edge of a trench* may produce a surcharge load which is too great for the trench sides to withstand. No materials or spoil should be placed within a metre of the trench and the height of spoil heaps near the trench should be limited to between 1 and 2 m.

Excavation is rarely in homogeneous ground. *Differing strata* may occur in relatively small vertical distances and inclusions of one type of soil may occur in another, e.g. pockets of sand may occur in a predominantly clay soil. Various sizes of boulder may occur, e.g. as in boulder clay, and sometimes there are pockets of water. Such mixtures considerably reduce the strength of the ground and limit its ability to stand unsupported for any length of time.

Previously excavated ground reduces the inherent strength of the adjacent ground – the backfill may have been of variable consistency and not too well consolidated. When previously excavated ground is close to or crosses the new trench collapse may well occur. Examples of this are as follows.

- When pipes cross a newly excavated trench the ground in the vicinity of a previous excavation will be very weak and particularly susceptible to failure.
- If a previously excavated trench exists approximately parallel to and within a meter of the new trench, the ground between them will be weak and very likely to collapse.
- A further example, not exactly of a previous excavation, is of that where hedges, bushes or trees are in close proximity to the line of the new trench whether or not they have been grubbed up before excavation starts. The action of the roots in taking moisture from the surrounding ground causes a breakdown in the strength of the soil.

The *sides of the trench may be struck by heavy loads* during the lowering of pipes or other materials into the trench.

When a trench is excavated close to and to a depth equal to or below the foundations of a building or structure the *surcharge load* may not only cause the collapse of the sides of the trench but also may well cause the collapse or partial collapse of the structure.

A further common cause of collapse is *insufficient and poor quality timbering* used to support the trench sides.

Prevention of trench collapse. The likelihood or otherwise of a trench collapsing can rarely be assessed by visual inspection. There is almost no ground that will not collapse given certain conditions (as illustrated above) and visual inspection cannot ensure that none of these will arise.

Sometimes, following a visual inspection, an assumption is made that the trench sides may well be self-supporting for a short period. This is a fallacy and should never be risked. Many fatal and serious accidents have occurred in unsupported trenches which have only been open for an hour or two – a substantial proportion have occurred during the day of the excavation. In 1971 Tomlinson[1] referred to experiments he had made relating to the stability of the vertical sides of a trench on a site where the soil consisted of soft to firm alluvial clay 1·5–2 m thick overlying soft to very soft peaty alluvial clay. The results showed that for various depths from 2·9 to 5·5 m, partial collapse occurred after 14–37 minutes and complete collapse after 23–49 minutes. The longest times, 37 minutes and 49 minutes, were for the shallowest excavation.

The ways to prevent collapse and make working in excavations safe are either by

- timbering or shoring, i.e. supporting the sides (the provision of physical support for each side of the trench)
- battering the sides to a safe angle.

Battering requires the sides of the excavation to be sloped no more than the angle of repose for the soil. This would obviously vary for the particular ground but in no cases should the angle be more than 45° to the horizontal and often much less, e.g. nearer 30°. A guide to the safe angle at which homogeneous soils will stand is given in *Timber in excavation.*[2]

Sometimes a combination of timbering and battering is used, usually in fairly deep trenches: for the first metre or so from the bottom of the trench, the sides are vertical and then they are battered to a safe angle. In this case the vertical sides must be supported as for timbering.

Supporting the sides of a trench may be carried out with timbering, H-piles, drag boxes or proprietary systems.

Conventional *timbering* comprises open poling or close sheeting supported by walings strutted across the trench – depending on the type and quality of the ground. Although this method is called timbering, sometimes steel trench sheets are used for poling boards and adjustable props rather than timber are used for struts.

H-piles are steel universal column sections pre-driven into the ground at intervals along the intended line of each side of the trench with their flanges parallel with each line. As excavation proceeds the exposed sides are supported by timbering or trench sheets laid horizontally behind the flanges of adjacent piles and wedged in position against the ground.

Instead of conventional timbering *drag boxes* may be used, which usually comprise steel framed boxes with strong side sheeting and struts. The two ends, top and bottom are open and the width is slightly less than that of the trench. The two sides extend to the full depth of the trench and are set on skids. They have to be strong enough to withstand any collapse of the trench sides. Men may then work at the bottom of the trench within the

Fig. 4.2. Drag boxes allow operatives to lay pipes in safety. The drag box is pulled along by the excavator as work proceeds. Best suited to open field construction in stable ground (courtesy Mabey Hire Limited).

boxes in safety. As trench bottom work proceeds, the boxes are pulled along, usually by the excavator, and the men remain inside within the protection of the side sheeting (Fig 4.2).

Various *proprietary systems* such as hydraulic struts and associated vertical poling members are in use for trenching (Fig. 4.3). Similarly, a hydraulic strut and waling frame system is used, in conjunction with ordinary trench sheets used as poling boards. A third type consists of specially formed vertical members with slots and incorporating struts. These are installed in facing pairs at intervals along the trench and heavy side plates are slid into the slots to the full depth of the trench (Fig. 4.4).

Before timbering is carried out or proprietary systems are used all material and equipment should be inspected to ensure that it is sound and suitable. With the installation of traditional timbering no work should be done outside the protection of the already constructed support. All installation work must be carried out from above or behind the completed work

Fig. 4.3. Vertical shoring system for use in ground not requiring close sheeting: the hydraulic cylinders are pressurised by a hand pump (courtesy Shorco)

(a)   (b)

Fig. 4.4. Standard trench boxes (a) courtesy Shorco (b) courtesy Mabey Hire Limited or from suitably strong prefabricated boxes which may be lowered into the trench and from which workmen can work in safety. Many fatal accidents have occurred to men installing the timbering when working no more than 0·5 m outside the previous support. The great advantage of the proprietary systems is that they can be installed from ground level without the need to enter the danger area of the trench.

There is a need to inspect and maintain the timbering because climatic conditions and work operations may affect its strength and stability, e.g. wedges may work loose. In Britain the CHSW Regulations require that any excavation must be inspected by a competent person

- before any person carries out work at the start of every shift and
- after any event likely to have affected the strength or stability of the excavation or any part thereof and
- after any accidental fall of rock or earth or other material.

Within 24 hours of making the above inspections a report has to be prepared in accordance with Schedule 8 of the CHSW Regulations.

Other hazards of trench work. Other hazards associated with trench work not involving soil collapse include

- workers being struck by falling materials
- workers being struck by falling rocks
- people falling into the trench
- unsafe means of access
- workers being struck by excavating machinery
- vehicles falling into the trench.

From time to time workers in trenches are *injured by falling materials* which have been lying about at ground level and are accidentally knocked into the trench or slip into it, moved perhaps by the vibration of passing vehicles. The materials may be pipes, bricks, adjustable props, trench sheets or timber, etc. intended for use in the trench. The accidents may be very severe (e.g. in a trench 2 m deep a pipe falling on to a man bending down may fracture his spine). Materials should not be left anywhere near the trench but should be properly stacked well clear. They should be taken to the trench only when ready for immediate use and then passed to those in the excavation.

Workers may be *struck by rocks* falling from the sides of the trench; the rocks may be loosened first during the excavation process and subsequently by vibration or from being struck when materials are lowered into the trench. This often applies when trenches are cut in existing roadways and the boulders in the sub-base are loosened or where rock blasting has been carried out. During timbering and before work in the trench begins, it is important that all loose rocks, boulders and stones, etc. should be identified and secured or removed. Also, that workers in the trench should wear safety helmets at all times.

Where people are likely to suffer *falls* of more than 2 m into a trench, rigid guard rails or barriers at least 910 mm high with a toe board and an intermediate rail to reduce any unprotected gap to 470 mm or less. The timbering may be extended above ground by at least 440 mm and the guard rail fixed to it. Alternatively, a properly constructed barrier fence parallel to and a short distance away from each edge of the trench should be erected. A post and guard rail type would be sufficient. In across-fields work where few people are likely to be in the vicinity a spoil heap with rope or brightly coloured bunting carried by posts at intervals on top of the spoil would be acceptable. However, even if the depth of the trench is less than 2 m it is advisable to take similar precautions to prevent anyone falling into it. Lighting or warning lights should be provided at night where appropriate. This is particularly important where the trenches are in public places.

Access may be required across the top of a trench at ground level. The access should be properly constructed and should be at least 430 mm wide for pedestrian use only – where it is intended to carry materials across (e.g. in a wheelbarrow) the width must be not less than 600 mm. Where the distance from the walkway to the trench bottom is greater than 2 m each side of the walkway must be provided with guard rails and toe boards.

Ladder access should be provided at reasonable intervals within the part of the trench where the sides have been adequately supported. On no account should a ladder that could be used be placed in an unsupported area of a trench. Ladders for access should be sound and properly constructed, should project at least 1·0 m above ground level and be tied to prevent slipping.

To prevent workers being *struck by excavating machinery* a safe method of working needs to be implemented. During the excavation cycle no person should be allowed in the vicinity of the machine. People in the trench should be well away from the face and those at ground level kept outside the slewing radius. When the excavator driver cannot see all parts of the jib and bucket during the excavation cycle or when the machine is used as a crane to lower materials (e.g. pipes) into the trench, an experienced banksman should be used to guide the driver and to ensure that people remain well clear of the operations being undertaken.

In the most common occurrence where *vehicles fall into a trench* small site dumpers – the front-end tipping type or other vehicles used for tipping materials (e.g. concrete) into the trench – either fail to stop at the edge or the ground gives way as they tip their load into the trench. Measures to prevent vehicles falling into excavations are required under the CHSW Regulations. The vehicles should be prevented from approaching too close to the edge by the use of adequate stops. These may be securely anchored timber baulks where the anchorage arrangement is made at a distance of twice the depth of the trench from the baulk near the edge. Proprietary stop block devices are also available. With this operation a banksman should always be employed to guide the driver.

Other reasons for this type of accident may simply be the inattention of the driver or too fast an approach. This points to the need for tighter supervision. It is essential that the driver is competent and should there be any doubt about his ability he should be given suitable training before being allowed to continue with the type of work in question.

Brake failure as a possible cause of accidents may be due to inadequate maintenance and this must be corrected. Sometimes a site vehicle such as a crane travelling about the site may approach too close to the edge of a trench, causing the ground to collapse. Where heavy vehicles are likely to travel close to trenches a barrier must be erected to prevent such approaches being possible.

Another type of vehicle-in-trench accident occurs in trenches alongside or in a working carriageway. With this type of trench it is essential to erect suitable warning signs well before, and at the approach to, the excavation. Also, suitable barriers and coning should be erected at the approaches.

Basements and wide excavations

Most of the hazards of trench work are also applicable to basements and wide excavations. However, supporting the sides of a deep and wide excavation may be safely carried out by the following methods.

Before excavation a diaphragm wall is sometimes formed around the perimeter of the site. This is designed to act as a retaining wall for the sides of the excavated basement. Following the construction of the wall excavation may then take place to the necessary reduced levels in perfect safety for the operatives, since the sides are prevented from collapsing. A similar system also in use is that of contiguous bored piles (Fig. 4.5) where a continuous line of bored piles to a depth below formation level is installed around the perimeter of the site. These are permanent supports and their use may have been decided at the design stage (*see also* p. 203).

Another method is by means of close boarding with timber, trench sheets or even sheet piles (Fig. 4.6). With timber runners or trench sheets these are installed progressively in a narrow excavation around the edges of the site. The runners are used in conjunction with walings at suitable vertical centres and the ground pressures transferred via the walings through a system of raking shores and thrust blocks at their lower ends, to the unexcavated ground. When the close boarding has been installed the rest of the excavation can be carried out in safety.

When constructing basements, deep excavations and shafts, it is particularly important to construct a solid fence around the site before excavation is started.

Shafts

With shaft construction the hazards are similar to those for trenches, i.e. materials falling into the shaft from ground level, earth or boulders falling

Fig. 4.5. Use of contiguous piles as permanent ground support before excavation at the Channel Tunnel portal through Castle Hill, Folkestone (courtesy Eurotunnel)

from the sides and people falling into the excavation. Support of the sides during shaft construction may be carried out with traditional timbering installed progressively with the shaft excavation. Where the timbering is properly carried out there is no danger to the workers in the shaft either from the collapse of the sides or from dislodged boulders. At the top, the timbering should be extended to form a fence so that people and materials are prevented from falling into the excavation. Where the timbers are not extended in this manner a close-boarded fence should be erected around the shaft top.

The use of proprietary shaft bracing in conjunction with steel sheet piles is illustrated in Fig. 4.6.

Shafts of larger diameter may be constructed with precast concrete lining rings installed as excavation proceeds, thus providing constant support to the sides. Where excavation is to proceed at a faster rate than lining installation some form of temporary support for the sides must be used. One method is to use sprayed concrete. This may introduce a health hazard from the overspray and rebound, so the operators must use full

Fig. 4.6. Use of proprietary shaft bracing with steel sheet piling (also shown is a mobile crane with outriggers extended and bearing on prepared foundations) (courtesy Mabey Hire Limited)

protective clothing and appropriate respiratory protective equipment (*see* Fig. 4.35).

Workers should only be conveyed down a shaft to their place of work by means of a man-riding hoist or suitable skip. Skips should be at least 1 m deep and provided with means to prevent occupants falling out and also with means to prevent the skip spinning or tipping. They should only be lowered under power by a suitable crane or winch.

Effective communications. With deep shafts the maintenance of an efficient communication system between the people working in the shaft and those at the top is important to ensure the safe control of the lowering and lifting of skips or other conveyed loads. Control is usually best vested in an experienced banksman at ground level who should ensure during lifting and lowering that people at the lower level in the shaft keep well away from the area directly below the load. Where a crane is used the crane driver must also be in the communication net and receive the necessary guidance from the banksman.

Dangerous atmospheres. Another type of hazard associated with shaft construction is that of dangerous atmospheres, which may result from

- oxygen deficiency
- carbon dioxide
- carbon monoxide
- nitrous fumes
- methane gas.

Steps to prevent risk of injury by anything liable cause asphyxiation are specifically required by the CHSW Regulations.

Tests for *oxygen deficiency* should be made before anyone enters the shaft at the beginning of each day. There should also be continuous testing while work is being carried out. In some types of ground a fall in the barometric pressure may cause a sudden change in the atmosphere to one severely deficient in oxygen.

Although an excess of *carbon dioxide* may be found in any ground it is particularly likely in chalk and limestone areas. Again the atmosphere should be tested at regular intervals for the presence of carbon dioxide.

Where internal combustion engines are used in the excavation the exhaust fumes will contain *carbon monoxide*. In these cases not only should the atmosphere be continuously tested but also an efficient ventilation system must be introduced in the shaft to provide a safe working atmosphere.

Nitrous fumes would occur if the use of explosives were resorted to during the construction process (e.g. if a band of very hard rock were encountered). Obviously during the firing and immediately afterwards no one will be in the shaft; nevertheless an adequate ventilation system must be provided to clear the nitrous fumes and to ensure a safe working atmosphere for the returning workers. The atmosphere must be tested before work starts to ensure that it is fit to breathe.

Methane gas may be encountered in or near coal measures and carbonaceous ground, and in or near refuse dumps. It is flammable and highly explosive when mixed with air at concentrations of 5–15% by volume (of methane). Constant monitoring of the atmosphere is essential when there is any possibility that methane may seep into the excavation. Where concentrations cannot be kept below 0·25% by volume, only explosion-protected electrical equipment together with an adequate system of ventilation must be used. However, if the concentration reaches 1·25% by volume all workers should be withdrawn from the workings.

Deep bored piles

With deep bored piles there are similarities to that of shaft construction in respect of the safety and health of the workforce. For inspection purposes, for bottom preparation such as cleaning out and sometimes for hand-digging extensions to under-reams, deep bored pile excavations must be entred. However, entry must never be made into boreholes with a diameter of less than 750 mm.

Those who descend boreholes and those who work in shafts or trenches encounter similar dangers. Materials or tools may fall down the shaft from ground level or workers may fall into the borehole from the surface. The sides may partially or even completely collapse and there is a likelihood that atmospheres are unfit for breathing. Where work has to be carried out in an under-reamed section there is the possibility of the roof collapsing.

Prevention of items from falling. All boreholes left unattended should be securely covered or have an adequate fence, reasonably close to the edge at ground level, to prevent people falling into them. No tools, materials or loose spoil should be allowed within a metre of the borehole at ground level and an upstand of at least 150 mm should be provided close to the edge. This may be carried out by allowing the lining tube to protrude above ground level. Another precaution to help ensure that tools do not fall down the shaft is for those people who descend the borehole or who work near it at the surface to be banned from carrying loose tools or equipment.

Entering a large diameter pile. Descent should only be made in specially designed safety cages for inspection purposes or in open muck skips fitted with an anti-spin device, and all lowering and lifting must be carried out under power control by means of a suitable crane or winch. While anyone is below ground the lifting appliance must be in the immediate charge of a competent operator and the power source kept running. Unless the operator of the lifting appliance can see the people working at the bottom of the borehole at all times a competent banksman must be employed at the top of the borehole to take charge of all lifting and lowering operations.

Before anyone enters the borehole to carry out work at the bottom the sides must be inspected by a competent person from a safety cage fitted with a roof strong enough to withstand falling material or other objects. It must be certified that the borehole is safe to enter (or otherwise). The person carrying out this inspection should remove any loose boulders or stones from the sides. Furthermore, if the borehole is to be unlined except at the top the inspections must be made at frequent intervals while people are working at the bottom.

Work in confined spaces is covered generally by the MHSW Regulations (*see* p. 194) and more specifically by the Confined Spaces Regulations (in preparation 1996).

No descent is to be allowed into an unlined borehole more than 12 hours after the start of excavation or more than 3 hours after completion – these periods may need to be shortened. Also, no person should remain in a borehole for more than 1 hour at a time.

Where strata are unstable, and in every borehole for the first metre, lining tubes must be used. At the completion of excavation and before concreting operations are begun particular care is needed with the extraction of the lining tubes to prevent any overloading of the lifting appliance. Where work is to be carried out within the under-reams the roof of the under-ream must be supported by traditional timbering.

Respiration. At least $1.5\,\text{m}^3$/min of respirable air must be provided at the bottom of the borehole for each person employed there. As for shafts, the dangerous gases likely to be encountered are methane, carbon dioxide, possibly carbon monoxide from vehicle exhausts at the surface, nitrous fumes if any blasting has taken place, and possibly ammonia and petrol fumes in contaminated ground from previous occupancy; but perhaps the major danger is from oxygen deficiency. Continuous monitoring for these gases and for oxygen deficiency is most important while anyone is underground and must in the first instance always be carried out by a person wearing appropriate breathing apparatus before anyone is allowed to enter.

Safety equipment and rescue. Those working in shafts and boreholes should wear safety harnesses and safety helmets at all times. Suitable rescue and first aid equipment including a stretcher, breathing apparatus, rescue lines and a winch (or crane) should be readily available for use near the top of each shaft (or borehole). While anyone is employed at the bottom an experienced banksman or rescuer must be on duty at the surface to observe that those at the bottom remain fully conscious and in no difficulty. Where the shaft or borehole is too deep for direct vision some form of communication must be established, such as field telephone, and regular checks made to ensure that those at the bottom are in no difficulty.

In an emergency the person at the surface is responsible for getting the workers to the surface as quickly as possible, after summoning additional help. Nobody should enter the shaft or borehole without breathing apparatus. Many multiple fatalities have occurred where rescuers have entered confined spaces without breathing apparatus to help colleagues, and have themselves been immediately overcome – especially where the emergency was due to oxygen deficiency.

Scaffolding

In the construction industry perhaps the most common way of providing a place to carry out work at a height is by means of a scaffold. The main hazards associated with scaffolding are the following.

- People falling from the working platforms.
- People below the working platform being struck by material falling or being thrown from it.
- The scaffold, or part of it, collapsing and throwing people from the working platform; or the collapsed structure crushing people under it or nearby at ground level.
- The collapsed scaffold causing damage to adjacent property or to the structure associated with the scaffold.

The CHSW Regulations require that all scaffolding should be erected, altered and dismantled only by experienced scaffolders under competent supervision. A test of competence is that the scaffolder has been trained under the Construction Industry Training Board scheme, and has received the appropriate certificate.

When it has been erected, to comply with the CHSW Regulations the scaffold must be inspected before being taken into use for the first time and

- after any substantial addition, dismantling or other alteration, and
- after any event likely to have affected its strength or stability, and
- at regular intervals not exceeding seven days since the last inspection.

The inspection must be carried out by a competent person whose duty it is then, within 24 hours, to complete a report as required by Schedule 8 of the CHSW Regulations.

Tube and fitting scaffolds

Tube and fitting scaffolds are constructed from loose tubes and a variety of fittings; the tubes are clamped together by fittings called couplers. Tubes and fittings are usually made of steel but may be of aluminium alloy. Steel and alloy tubes and fittings should never be used on the same scaffold.

The design and layout of tube and fitting scaffolding are fully described in BS 5973: 1993 *Code of practice for access and working scaffolds and special scaffold structures in steel.*[3] The most common types of tube and fitting scaffold are the putlog or bricklayer's scaffold and the tied independent scaffold.

Putlog scaffolds. The putlog scaffold is mostly used in connection with the construction and maintenance of masonry and brickwork structures and is sometimes referred to as the bricklayer's scaffold. It consists of a single line of standards (uprights) parallel to the face of the building and set about one metre away from it (to accommodate a platform four or five boards wide). Horizontal tubes called ledgers are connected to the standards at vertical intervals of approximately 1·35 m by right-angle (load-bearing) couplers. Putlog tubes span the distance between the ledgers and the building and support the working platform. The flattened blade or spade end of the putlog sits on the brickwork and is built in as the work progresses. The outer end should be fastened to the ledgers by putlog couplers to prevent movement. The decking platform usually consists of scaffold boards 225 mm wide × 38 mm thick and only one working platform may be in use at any one time.

Tied independent scaffolds. An example of a good tied independent scaffold is shown in Fig. 4.7. It consists of a double row of standards each parallel to the face of the associated structure. The inside row should be as close as possible to the structure and the distance between the lines of uprights should accommodate a platform of four or five boards. Sometimes the first row of standards is set away from the building face to allow for one board to be positioned between the building and the line of standards. The lift height (the distance between the ledgers) may vary between 1·8 m and 2·7 m but is generally about 2·0 m. The boards forming the working

Fig. 4.7. A good example of a tied independent scaffold. Working platforms pre-1996 have unprotected gaps greater than 470 mm and do not comply with the CHSW Regulations (courtesy HSE)

platforms rest on tubes called transoms which span the distance between the ledgers. These should extend inwards to touch the building face.

Falls from scaffolding

People sometimes lose their balance at working platform level because they slip, trip, stumble or on very rare occasions are taken ill. The presence of toe boards and guard rails on a scaffold will generally prevent a fall to a lower level should a worker lose balance. Toe boards and guard rails should be fitted to the outer side and ends of the working platform and be so secured as to prevent their outward movement – this is best carried out by positioning them on the inside of the standards. The toe boards (e.g. scaffold boards on edge) should be at least 150 mm high from the decking (i.e. the working platform) and constructed so that no gap exists between the decking and the lower end of the toe board. Guard rails usually consist of steel scaffold tube erected horizontally and fixed to the standards by

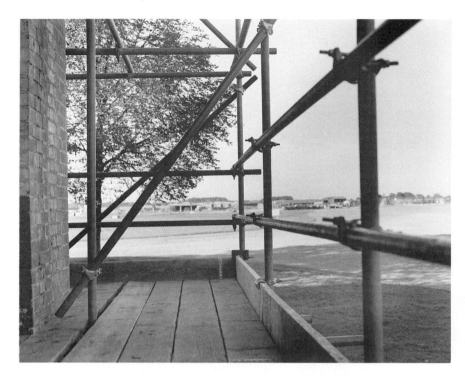

Fig. 4.8. A well constructed working platform on a tied independent scaffold (courtesy CITB)

means of right-angle (load-bearing) steel couplers. They should be located at least 910 mm above the decking with an intermediate rail or barrier so as to have no gap more than 470 mm. Examples of good and bad working platforms are shown in Figs 4.8 and 4.9.

A hazard which might cause a person to slip or trip should of course be identified and eliminated to prevent the possibility of an accident. Areas that are slippery after spillages should be immediately cleaned and sanded as necessary. Chippings (e.g. from bricks) and dust from materials should be swept up and not allowed to accumulate.

Materials deposited on the scaffold platform should be neatly stacked and a clear passageway should be maintained between the materials and the edge of the platform. The CHSW Regulations require that working platforms shall not be less than 600 mm wide. For the transport of materials along the scaffold platform neither the guard rails nor the toe boards should be removed.

The surface of the platform should be even and without excessive sag caused by the self-weight of the boards, by people moving along it or by the transport of materials (e.g. in barrows). All the boards forming the platform should therefore be of the same thickness and this should be

Fig. 4.9. An unacceptable scaffold at working platform level with no guard rails or toe boards, a large gap between boards and inadequate passage width (courtesy HSE)

chosen having regard to the intended loading and span. The recommended maximum span for the most commonly used scaffold boards where the platform decking consists of a single layer of boards and the distributed loading is not intended to exceed 3 kN/m^2 is 1 m for 32 mm board, 1·5 m for 38 mm board and 2·6 m for 50 mm board. Where a higher loading figure is required the size and spacing of the boards will need to be specially designed.

In addition, each board should be supported by at least three transoms or putlogs. Their ends usually abut and one transom or putlog is required near each end of a board. The transoms or putlogs should be spaced so that the board ends do not overhang by more than four times the thickness of the board, e.g. 150 mm for a 38 mm board. This will prevent the possibility of the board tipping under load and throwing someone off

Fig. 4.10. Spacing of transoms or putlogs: (a) at a joint; (b) at the end of a run

balance, e.g. someone walking on to the cantilever or pushing a loaded wheelbarrow with the wheel running on to the area. Also, to ensure that through vibration or other movement a board does not slip off its end support the overhang should be not less than 50 mm (Fig. 4.10).

However, sometimes the boards overlap at the joint and then only one supporting transom or putlog is required. In this case it is necessary to provide bevelled pieces, suitably secured, to overcome the tripping hazard and to facilitate the passage of wheelbarrows (Fig. 4.11).

Workers may also fall through the scaffold platform if a large enough gap exists or is created. To prevent this the boards should be laid with the minimum of space between adjacent boards. Also, boards should not be removed from working platforms during the life of the scaffold except when they need to be replaced, in which case the replacement should be immediate, or that part of the scaffold that is affected should be fenced off to prevent it being used until the replacement has been carried out.

Falling materials

Where there is any possibility of people below being struck by material or tools falling through a gap in the working platform the boards must be laid close boarded, i.e. side by side and end to end without space between the edges of adjacent boards.

As well as preventing people slipping from the outer edges, the toe board has another function, i.e. preventing material and tools falling. Where materials are stored on the scaffold to a height above the 150 mm high toe board either a higher toe board (e.g. two or more above each other) is necessary, or preferably the space between the upper guard rail and the top of the toe board should be filled in with, for example, stout wire mesh frames suspended from the guard rail. They are often referred to as brick

Fig. 4.11. Spacing of a transom or putlog where board overlaps at a joint

guards and it is of course essential that their outward movement be prevented (*see* Fig. 4.15).

Material, tools or other equipment should never be thrown from the platform but should always be lowered under control in a suitable skip by means of a gin wheel, scaffold jib crane or hoist. This also applies to any component of the scaffold during the dismantling phase of the scaffold structure.

Material or equipment should also be conveyed from a lower level to the working platform by similar means. When lifting or lowering by means of gin wheels or scaffold jib crane it is necessary to ensure that the skip is not accidentally detached from the suspension hook during its travel – only safety hooks should be used. Also, by careful control and visual monitoring of the path of the load throughout its travel, contact with the scaffold and subsequent tipping of the skip should be avoided.

In the construction of scaffolds other methods of preventing people being struck by falling objects include total enclosure – the suspension of sheeting (e.g. canvas sheet) at all the outer edges or the use of protection fans projecting from the scaffold. The decking for fans may be of waterproof sheeting, close-boarded timber or safety netting – where the latter is used it is important for material netting to be laid over the heavier safety netting to prevent small objects falling through the larger mesh.

Scaffold collapse

Partial or total collapse of putlog and tied independent scaffolds may cause major injury or death both to users and to those below the scaffold and considerable damage to adjacent property and to the structure associated with the scaffold.

The scaffolds should be erected in accordance with British Standard BS 5973 (1993) and failure to do so in any respect might cause or contribute to collapse. Some examples of poor construction commonly found are

- standards erected out of plumb
- putlog clips used in places where load-bearing couplers should be used
- fittings incorrectly (loosely) tightened
- excessive spacing of standards
- inadequate bracing
- poor foundations, for example standards founded on uncompacted ground, baseplates omitted (only acceptable when standards are pitched on concrete or steel of adequate thickness), lack of suitable sole plates.

However, the main cause of collapse falls into one of two categories: insufficient ties (Fig. 4.12) and overloading.

Ties. Ties are required to attach the scaffold structure rigidly to the

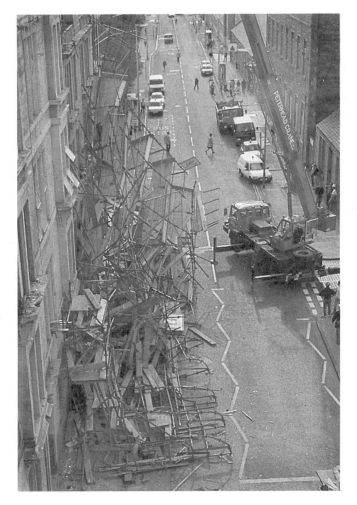

Fig. 4.12. Collapse of tied independent scaffold into a busy street due to insufficient ties and bracing (courtesy HSE)

associated building and they should prevent movement of the framework towards and away from the building. Rakers may be used instead of ties if required but the rakers must be securely fixed and on firm foundations. There are two classes of tie, movable and fixed. The use of movable ties is necessary to allow following trades (plasterers, glaziers, etc.) temporarily to remove a tie (one at a time), in order to carry out their work, but the ties should be replaced as soon as possible. Fixed ties must not be removed. It is essential for the stability of the scaffold that the rules in respect of ties and their spacing as required by BS 5973: 1993[3] are complied with.

Overloading. When a scaffold is in use to facilitate construction, maintenance or demolition of a building or structure, overloading may result from the stacking and storage of material associated with the works on the platform. Generally well-constructed scaffolds will accommodate a degree of overloading but collapse may occur with relatively small overloads where the number and spacing of ties is inadequate and/or one or more of the faults listed on p. 71 are present.

Depositing material on to the platform(s) from a crane or fork-lift truck requires a great deal of care to prevent a heavy landing, and thus a high live load, for which the scaffold is not designed. This is best dealt with by a banksman at landing level directing the crane or fork-lift truck driver regarding the movement of the load. Similarly, a banksman is vitally important for the safe conveyance of loads by crane, and, where the driver does not have a clear view of the whole travel of the load, for preventing collision with the scaffold.

The scaffold platform is sometimes used to stack material from demolition operations and care must be exercised to prevent large pieces of masonry or other material dropping on to the working platform. Also, as the building is progressively reduced in size, it is necessary, especially at the lowest levels, to ensure that the required numbers of ties are maintained or that the scaffold is stabilised by means of rakers.

It is also fairly common for collapse or partial collapse to occur during dismantling. This usually happens when material (tubes, boards and couplers) from the higher areas of the scaffold is stacked at the first or second lift level to await a lorry to take it away. The material is stacked over a relatively small area of the scaffold, resulting in a massive overload which the remaining structure is simply unable to carry. Preferably the material should be lowered to the ground and stacked to await removal, or if this is not possible and it has to be stored at the lower levels, the scaffold should first be adequately strengthened. Also, dismantling should be evenly carried out in horizontal layers to the full width of the scaffold and not in vertical sections.

Other causes. Other causes of partial collapse are

- in putlog scaffolds, the dislodgement of the spade end of a putlog causing the platform to fall
- the failure of a scaffold board(s)
- collision of motor vehicles with the structure.

In the case of putlog scaffolds the spade ends of the putlogs must rest on or in the brickwork to a depth of at least 75 mm. This will prevent accidental displacement caused by vibration and other scaffold movement.

Scaffold boards should be carefully examined immediately before their use in a working platform. They should be free from unacceptable faults such as large knots, knot clusters or large splits. However, perhaps the most common reason for the failure of scaffold boards is reduced strength owing to a compression fracture (of the wood grain) from previous careless use. The fractures are often caused when boards are dropped from a height, especially if they land on a sharp edge such as the edge of a concrete raft or on the edges of broken bricks on the site, and when boards are left lying about the site and vehicles run over them. Compression fractures greatly reduce the strength of a board and are very difficult to identify even by experienced examiners. Also, boards should never exceed the spans recommended above (*see* p. 69).

Unless the scaffold is constructed in an area where little or no traffic is anticipated the standards should be protected at their base by fenders (e.g. large timber baulks laid around the outer edges of the structure) to prevent vehicles colliding with the structure.

Proprietary scaffolds

Independent tied scaffolds may also be constructed from proprietary prefabricated units of which several systems are available. The safety precautions for tube and fitting scaffolds apply also to proprietary systems; they should be plumb and on firm bases, with all joints secure, and they should not be overloaded. Although proprietary scaffolds may not require the same standard of bracing as loose tube and fitting scaffolds they all require the same standard of tying. When cranes are working in the vicinity, care should be taken to prevent snagging the scaffold because, in some types of proprietary system, the joints have no resistance to an upward thrust.

Tower scaffolds

Tower scaffolds may be of the stationary or mobile type. They are usually rectangular or square towers constructed from loose steel tubes and fittings or proprietary types comprising prefabricated frame sections with special joint arrangements making assembly simple and speedy (Fig. 4.13). They are also often made from aluminium alloy as well as steel (Fig. 4.14). In all cases they provide one working platform. To prevent people falling from the working platform, guard rails and toe boards are required as for the common putlog or independent tied scaffolds. Also, similar transom spacing, board spans and board qualities are required.

With mobile towers a common accident is for the scaffold to move inadvertently while someone is on the platform, and a wheel(s) hits an obstruction with the result that either the scaffold overturns or tips sufficiently violently to throw the person(s) from the platform. To prevent this it is essential that all the wheels (castors) are provided with efficient locking devices and that these are engaged while people are using the platform. Also, the wheels must be positively secured to the uprights to prevent them falling off. A similar type of accident results from the tower being deliberately moved at the base (to a new working area) while the platform remains occupied. This action should never be carried out – no one should occupy the platform while the tower is being moved. Also, the towers should only be used on firm and reasonably level ground.

Fig. 4.13. Mobile work platform used for frame erection at Villa Park. Note use of an all terrain mobile suspension system set up on the terracing (courtesy Hewden Stuart Limited)

Towers may also overturn where an excessive horizontal force is applied – usually at the working platform by the user, but the force may be due to wind load in a location outside a building. A general rule for ensuring stability for most operations is that the height to smallest base dimension ratio must not exceed the following in respect of steel towers

- 4:1 for stationary towers used inside a building
- 3·5:1 for stationary towers used outside
- 3·5:1 for mobile towers used inside a building
- 3:1 for mobile towers used outside.

The height is measured from the floor to the working platform level and the least base dimension is the width from centre to centre, of the shortest side. Where outriggers are used the width over the shortest side of the outrigger base is used. Where aluminium towers are used the above ratios would be reduced.

A particularly common accident with tower scaffolds occurs where the

Fig. 4.14.   Alloy access towers are easy to transport and erect and are ideal for many maintenance tasks (courtesy Leada Acrow)

operative(s) wishes to reach a work area substantially higher than the base-to-height ratio for the platform level will allow. In order to overcome this difficulty, frequently operatives erect a ladder on the platform, usually with the foot of the stiles hard against the toe board. Sometimes one person will 'foot' the ladder while the other starts to climb to the working area. Inevitably the horizontal reactive force at the bottom of the ladder due to the weight of the ladder and the person (plus tools or material) climbing produces a much greater overturning moment than the scaffold's natural (self-weight) righting moment and the scaffold overturns, usually with disastrous results. This method of work – placing a ladder on the working platform of a tower scaffold – should never be used.

Birdcage scaffolds

Birdcage scaffolds are intended for light work such as decorating, plastering or fixing lighting, or for light maintenance work within buildings with large floor areas such as factories, cinemas and churches. They consist of parallel rows of standards at regular intervals evenly spaced in both directions – not exceeding 2·5 m centres. They are connected using right-angle couplers to ledgers and transoms at each lift height, which is generally about 2 m. The top lift only is boarded out to form the working platform, and here the transom spacing must be such that the span of the boards is appropriate to their thickness (*see* p. 69). Very often birdcage scaffolds are erected to cover the whole of the floor area of the part of the building where they are used. If the platform is then constructed so that no, or only a very small, gap exists at the edges, toe boards and guard rails are unnecessary but if the gap is over 25 mm and people below are likely to be struck by falling material then toe boards should be used. If the gap is such that someone might fall through the opening, guard rails are also required.

The most common accident with this type of scaffold is either complete or partial collapse. It is usually caused by very excessive overloading coupled with various elements of poor construction (e.g. standards not plumb, use of incorrect, non-load-bearing couplers). However, the most common fault is the lack of adequate bracing in two directions to prevent excessive sway and ultimate collapse. Additional measures to prevent collapse are adequate tying and the extension of the transoms and ledgers to butt against the sides of the building (around all the sides of the scaffold). Ties may be of the box type around suitable columns within the building or reveal at window openings. One such restraint point (either a tie or a butting tube) should be provided for each 40 m^2 of vertical face and no standard should be more than six tubes away from a restraint point.

Slung scaffolds

Slung scaffolds are working platforms suspended by means of lifting gear, wire ropes, chains or rigid members (usually scaffold tubes) at a fixed distance below structural elements of a building (e.g. roof trusses or other suitable overhead load-bearing members) and not provided with means of raising or lowering by a lifting appliance. The platform is usually supported on scaffold tube ledgers and transoms which should be connected by right-angle couplers. The transoms should be spaced so that the span of the boards does not exceed that appropriate for their thickness. As for all scaffolds the platform should be close boarded and provided with suitable guard rails and toe boards at the outer edges. Where two transoms are used to support the joint of two boards their distance apart should allow for an overhang of at least 50 mm but not greater than four times the thickness of each board. This type of scaffold should always be specially designed and the suspension points for the platform supports should not exceed 2·5 m centres.

A major hazard particular to this type of scaffold is the failure of the suspension arrangement, either the suspension gear or the anchorage. In order to avoid or at least substantially to reduce the likelihood of an accident it is necessary to adhere to the following points.

- Where wire ropes or chains are used they should be adequate for the purpose and should have been tested. Scaffold lashings should never be used.
- A sufficient number of suspension points should be provided to prevent the platform tipping if one of the suspension ropes etc. fails (e.g. at least five points are needed for a circular platform and six for a rectangular platform).
- The wire rope etc. suspenders should be kept vertical as far as possible – if they are not this must be taken into account in the design.
- The wire ropes etc. should be secured to the platform ledgers as close as possible to their junctions with the transoms.
- At the platform end, wire rope suspenders should be secured to the ledgers by two full turns around the ledger and the end held back to the suspension wire by three bulldog grips, or by means of a suitable shackle when the end of the rope terminates in a tested eye.
- At its upper end the rope etc. should similarly take two round turns of the suspension member and the tail should be secured to the running rope by three bulldog grips.
- Where the wire rope at the upper suspension point is taken around sharp edges, the rope should be protected at all such high stress raising points by means of appropriate packing.
- The suspension ropes should have a factor of safety against failure of at least 6.
- The point of suspension and member from which the scaffold is suspended (e.g. members of a roof truss) should have a factor of safety against failure under total maximum load conditions of at least 2.
- Where suspension chains are used both ends should be secured by shackles.
- Where scaffold tubes are used as suspenders only right-angle couplers should be used and check couplers should be incorporated to protect against possible slippage.
- Vertical axial joints in tubes should not rely solely on sleeve couplers or joint pins. The joint should be made with a lapping tube and a suitable number of right-angle couplers.
- Material and equipment should be spread evenly over the platform and heavy point loads should be avoided. Workmen using the platforms should not congregate in any one bay.
- Guy ropes should be used to minimise platform sway.

Suspended scaffolds

Suspended scaffolds are sometimes referred to as cradles and are used for light work such as painting and repair work on the outside of buildings. The permanent type of power-operated suspended scaffold provided on some buildings is not dealt with here.

The cradles are suspended by fibre or wire ropes and are capable of being raised or lowered by manual or power operation. They may be of fixed type or be capable of travelling horizontally on a runway as well as being raised and lowered. The most common accident occurrence is the scaffold falling because of failure of the suspension ropes or the outriggers or, with the travelling type, running off the end of the track. Suspension ropes should not be used if they have suffered undue wear or damage that affects their strength. They should be properly reeved on to winding drums or winches (where used) over all pulleys and be correctly secured at anchorage points. Where the winches are of the drum type the length of the suspension rope should be such that at the cradle's lowest point there should be at least two dead turns of rope on the drum – to alert the operatives the last few metres of rope should be distinctly coloured. The rope anchorage on the drum should also be capable of sustaining the rated load of the winch. Also, winches used in connection with the scaffolds should be provided with a brake which comes into operation when the operating handle is released.

Outriggers may be of steel joists, timber poles or sometimes suitably braced steel scaffold tubes. They should be

- of sufficient strength for the maximum load to which they will be subjected
- project beyond the face of the building just far enough to ensure that the suspended cradle will not foul on any parts of the building during its upward or downward travel
- either bolted down at their inner (tail) end or securely counter-weighted to prevent overturning under maximum load conditions; the total load includes dead load, live load and an impact factor (25% for power-operated and 10% for manually operated scaffolds); the counterweights where used must be properly attached to the tail end of the outriggers – loose bags of sand or rolls of roofing felt are not suitable methods of counterweighting; the counterweighting should provide a resisting moment of at least three times the maximum overturning moment
- kept as nearly horizontal as possible but never with the outer end lower than the tail end
- spaced reasonably close together but suitable for the maximum loads to be carried.

Where a runway is incorporated to facilitate sideways (horizontal) movement of the cradle it must be securely lashed or shackled to the outriggers. The jockey wheels are prevented from running off the end of the runway beam by securely fixed end stops at each end of the beam.

From 1996, non-power operated suspended scaffolds became subject to the Construction (Lifting Operations) Regulations 1961 (*see* p. 19).

Ladders

Access to many scaffolds is provided by ladders – usually timber ladders. Also, the ladders themselves are very often used as working platforms for the performance of light work, e.g. painting or inspection. Regrettably each year many serious accidents result from the misuse of ladders. Accidents occur because

- ladders slip when users are climbing or working from them; over 50% of all ladder accidents may be attributed to this cause
- users slip or miss their footing while climbing
- users overbalance when carrying materials or tools
- when defective ladders are used, they fracture under the weight of the user.

The legal requirements for ladders are given in the CHSW Regulations. When used for access, ladders should be securely tied near their upper end – preferably by each stile to the ledger at platform level. They should also stand evenly on their stiles, on a level and firm surface – never on loose bricks and never carried by a rung resting on a plank across a depression (e.g. a trench) in the ground. The rungs and users' footwear should be kept clean and free from slippery mud, for example. A particularly long ladder should be secured to prevent undue sway. At scaffold platform level the ladder must extend at least 1 m above the landing place (platform level) to provide a handhold while the user steps on or off. If this is not possible a suitable alternative handhold must be provided. There should also be sufficient space at each rung to provide a good foothold.

When carrying materials and tools invariably the user is climbing with only one hand on the stile. Alternative methods of lifting the material and tools should always be considered.

Where work is to be performed from the ladder, where the ladder cannot be tied at the top or adequately secured at the bottom to prevent movement, the ladder should be footed. The ladder should extend beyond the place of landing by at least 1 m unless adequate handhold is available and there must be sufficient space at each rung to provide adequate foothold. Many accidents are caused by over-reaching, causing the ladder to slip sideways – if the ladder is not secured at the top the user must exercise great caution when reaching sideways. Generally, ladders should be placed at an angle of 75° to the horizontal, i.e. 1 m out for every 4 m of vertical height when used for both access and working purposes (Fig. 4.15).

Timber ladders should be carefully handled – as with scaffold boards, they should not be dropped or thrown from heights and they should not be left on the ground where vehicles may drive over them. They should be stored in ladder racks and not hung by their stiles. Makeshift ladders should not be used, particularly those where the rungs depend solely for support on nails, spikes or similar fixings.

Fig. 4.15. Recommended practice for ladders used for access to scaffolds

Ladders with missing or defective rungs should never be used and where wire tie rods are used the ladder should be erected so that the rods are below the rungs. Metal ladders and timber ladders with metal reinforcements should never be used in the vicinity of electrical apparatus or equipment.

Ladders should be regularly inspected to identify defects and timber ladders should not be painted since this may hide defects – transparent varnish or linseed oil may be used as a preservative.

Falsework

Falsework is defined as a temporary structure used to support a permanent structure during its construction and until it becomes self-supporting. Falsework may be required to support steel and timber frameworks and masonry arches as well as in situ and precast concrete construction. The design and construction of falsework is covered by BS 5975 : 1982 *Code of Practice for Falsework*[4]

Falsework is often constructed from conventional loose steel scaffold tubes and fittings together with adjustable telescopic props and forkheads (sometimes flat heads) which carry the main timber bearers on which the formwork or deck is built. Secondary beams may be interposed between the main bearers and the formwork. Prefabricated steel frames, military trestles, towers, steel girders and standard steel sections are employed for more specialised work and generally where the load to be supported is particularly heavy.

The total or even partial collapse of falsework may lead to serious accidents for those on the structure who are thrown from or fall from their place of work.

More often than not collapse occurs when the structure is being loaded, e.g. during the placing of concrete when the number of workers involved in the operations is high. Where other work is being undertaken below the loading level the people involved are also at considerable risk from falling materials, i.e. both from the falsework and from the supported permanent works.

Accidents also occur without the falsework collapsing. These are the result of people slipping or falling or both from the structure. In addition to their major supporting roles the falsework and the formwork provide a place of work, and normal edge protection requirements are necessary at all open sides and holes in the decking, as well as properly constructed access to all places where work is to be carried out (Fig. 4.16). The collapse of falsework can be attributed to inadequate design, poor construction or a combination of the two.

Design

All except the very simplest falsework should be designed in accordance with recognised engineering principles and clear drawings, sketches and specifications should be produced for use by the site personnel so that the structure can be built in accordance with the designer's intentions.

It should be noted that designers of temporary works are subject to the full impositions of the CDM Regulations concerning the duties of designers and to the co-ordinating activities of a project's planning supervisor (*see* p. 24).

For fairly simple repetitive applications standard solutions may be used. Some examples are given in BS 5975.[4] However, when the standard solu-

Fig. 4.16. Safe working platforms on falsework for the bridge piers for the Queen Elizabeth II Bridge, Dartford (courtesy Trafalgar House Construction)

tions are used they must be developed by means of proper engineering design methods and the users on site must always be provided with drawings or sketches and if necessary a specification to ensure that construction is carried out in accordance with the standard solution.

Deficiencies in falsework design may arise from

- failure to estimate the type and extent of the loading correctly
- inadequate foundations
- incorrect choice or use of materials
- lack of provision for lateral stability.

Failures may also occur because of design errors which, given the temporary nature of falsework, are prone to occur if checking is neither thorough nor systematic.

Loading. Falsework collapse has occurred because the designer has failed to take into account all the loads applied to the structure. The self-weight of the falsework and formwork must include any ancillary temporary works connected to them, such as hoist towers, loading platforms and access ramps. Imposed loads must also be taken into account when they apply, such as

- the self-weight of the permanent works to be supported
- dead loads from static plant such as welding sets and compressors

- dead load due to stored materials, i.e. reinforcement or timber
- impact in placing the permanent works (e.g. units lowered by crane on to the supporting structure); where wet concrete is being placed, the impact when tipped from a barrow, a dumper, a skip or a concrete pump pipeline
- where wet concrete is to be pumped through a pipeline to the placement position both the dead load of the pipeline and the horizontal load due to the impulse effect
- the surge effect of wet concrete due to the use of poker vibrators
- those due to the hydrostatic-type pressure of fresh wet concrete on the faces of the formwork; the effect is more complex if the soffit is sloping
- those due to the expansion of concrete during hydration
- rolling loads from moving plant such as dumpers and barrows, including the effect of braking with mechanically propelled vehicles
- moving load due to operatives and an allowance for small plant and tools that they use in the course of their work
- those due to post-tensioning of the permanent works
- wind load on the falsework, the formwork and the completed part of the permanent works carried by the falsework; also, as necessary a snow load
- where the falsework is founded in a water course the dynamic action of the flowing water on the supports; consideration should also be given to the impact of floating debris on the supports and to the load imposed by the flowing water impinging on debris trapped by the supports; sometimes the falsework is erected in a dry watercourse but flash floods should be anticipated and the above load considerations applied.

Foundations. Many falsework structures are founded on railway sleepers or other types of sole plate and the designer should specify the need to ensure that the underlying ground is well consolidated and that any voids are filled with concrete.

The stability of the foundations may be seriously affected by scour from storm-water. If this is likely there may be a need to provide cut-off drains and to place the sole plates on surface blinding. Non-cohesive soils are particularly vulnerable to the effect of water and a high water table may considerably reduce the ground bearing capacity. Where uncompacted they are also susceptible to consolidation under the effect of vibration, e.g. of the wet concrete, which may cause differential settlement. When falsework is erected in dry watercourses a more substantial concrete foundation may be necessary to prevent scour caused by flash floods.

In the case of timber sole plates, the designer should anticipate and allow for an eccentricity of 25 mm for the uprights on the centre line of the sole plates. Where the sole plates are founded on sloping ground the falsework uprights and base plates should be built on wedge-shaped timber packing with cleats immediately below the wedges, and wedges and cleats properly secured to the sole plates. Thrust blocks at the base of the slope are required to prevent the movement of the sole plates down the slope.

At the design stage it is necessary to take into account the intended loading pattern to avoid differential settlement of the foundations. Such settlement would cause a redistribution of the loads on the falsework structure and might lead to an overload.

Where the falsework is supported on the permanent works such as in multi-storey building construction the falsework foundation design should be based on the strength of that part of the permanent structure that is intended to carry the falsework. The strength will be that at the time of loading and not necessarily the ultimate strength.

Materials. For falsework applications the materials used are often not new, but may have been used many times, especially for repetitive work of a similar nature. Materials may have been used for other types of work and been subject to abuse by overloading, rough handling or poor maintenance. The designer must be aware of the limitations of the material to be used for falsework under consideration and should adjust his permissible stresses accordingly. A particularly common reason for the need to reduce allowable stresses is the reduction in cross-sectional area of steel components due to the effect of corrosion. Where second-hand material is to be used it should first be inspected. Where cross-sections of components are severely reduced by corrosion or by previously drilled holes or where they are excessively damaged, bent or twisted, they should be rejected.

Stability. One of the main causes of collapse is the lack of sufficient lateral stability. It is important to prevent initial movement of a falsework system towards a collapse condition. The forces required to prevent initial sway are relatively small but when sway is induced, the inertia of the supported load and increasing eccentricities often make collapse inevitable.

The various horizontal forces due to some of the loadings identified under loading must be resisted by an effective bracing or restraint system. In addition to any bracing system, the designer should where possible require the falsework to be tied back to completed parts of the permanent structure or be firmly anchored to a suitable foundation.

Where an accurate assessment of the horizontal loads cannot be made horizontal forces equivalent to 2·5% of the applied vertical loads, considered as acting at the points of contact between the vertical loads and the supporting falsework, must be catered for as advised by BS 5975.[4]

Forces on mobile falsework. Some falsework or parts of falsework are moved on skids or wheels to another position for reuse. The designer must take into account the additional forces that may be imposed on the structure because of this. The need to overcome uneven track or ground conditions may subject the structure to excessive twisting or bending and this may also occur when manoeuvring it into its next position.

Buckling. It is important to detail all points of load transfer (e.g. from timber bearer to forkhead) in the knowledge that slight misalignments in erection can cause buckling and progressive failure.

When steel beams form part of the structure it is important that at all load transfer points the webs are checked for their ability to resist buckling. Where calculations are not produced which satisfy this requirement web-stiffeners must be provided at all such points. Failure to do so will almost certainly lead to collapse.

Traffic access. It is quite often necessary to allow vehicular traffic to pass through the falsework structure, e.g. where it straddles a roadway. To prevent vehicles colliding with the framework, a suitable fendering system should be provided. This should include a lead-in phase and not just comprise the immediate protection of the falsework.

Design brief. To produce a suitable solution for the falsework require-ment the designer must be provided with a brief. This will include such information as

- the self-weight of the permanent structure to be supported
- the falsework components to be used
- the access requirements to and through the structure
- data from the soils investigation
- information in respect of watercourses and sources
- the method of construction and the loading sequence
- the plant to be used and the method of placement
- the storage requirements e.g. at deck level
- in multi-storey construction the load-bearing capacity of the floors.

Construction

It is of paramount importance that the construction of the falsework follow exactly that shown on the drawings and in the specification produced by the designer. No alteration whatsoever should be made without consulta-tion with and approval of the person responsible for the design. It is also important that activities which have not been identified in the design brief but which may affect the stability of the falsework should not be carried out without clearance from the designer, e.g. excavations in the vicinity of the structure.

Failure to observe basic details of good construction will often be the cause of a partial or total collapse. Some examples of these basic details are now given.

Materials. All materials should be inspected before use and any not of the required quality should be removed. Where scaffolding components are to be incorporated only load-bearing couplers should be used, and putlog clips should be taken away from the storage area.

Foundations. The area beneath the sole plates should be well consoli-dated and where required the blinding should be laid or the concrete foundation constructed to the specified depth and area. The vertical members of the structure should be located centrally (laterally) on the sole plates of base grillage and none should be within 300 mm of the ends.

Adjustable steel props and forkheads. Only props of the correct size should be used. Load is transferred from one part of the prop to the other through a high-tensile pin and this should never be replaced by a makeshift pin, e.g. a reinforcement bar.

Props should not be bent and should be erected as plumb as possible but never more than 1½° out of the vertical, i.e. not more than 25 mm in 1 m. They should be positioned centrally underneath the member to be supported – any eccentricity should never be greater than 25 mm. At the top of the prop the supported timber beam should be square to the plate and secured in position by nails.

With adjustable forkheads the supported bearers should be positioned to avoid any eccentricity greater than 25 mm on the forkhead. They should be rotated to centralise the bearers over the tubular stem and matched wedges should be used to maintain the bearers in place. In addition the wedges should be secured by nails through the holes in the sides of the forkheads. The bearers should preferably be butt jointed and the joint should be within 15 mm of the centre. Where a beam terminates in a forkhead it should extend at least 50 mm beyond the centre point.

Where lapped joints are used for the bearers they should not span more than three supporters. Over-extension of the screw jack supporting the forkhead must be avoided. This must not be more than 300 mm without lacing (horizontal bracing) at right angles to the bearers, as near to the top as practicable, and with diagonal bracing in both directions at every sixth jack. Where adjustable base plates are incorporated similar bracing arrangements are required where the jack extension exceeds 300 mm.

Support components. The uprights should be centralised on the sole plates as for props and they should be erected plumb within a tolerance of 15 mm over a height of 2 m, subject to a maximum of 25 mm over the full height of the structure. Only load-bearing couplers should be used and these must be properly tightened.

For fabricated steel floor centres the correct spacing must be ensured and full bearing achieved at each end of individual centres on the end bearers. They have little or no resistance to horizontal forces and an adequate bracing system is essential before concreting is carried out.

Loading. Where precast concrete or prefabricated units are lowered on to the falsework a banksman should control the lowering operation and ensure that shock loading is prevented as far as possible. Also, the dragging of units into final positions should be avoided.

On the deck, materials should only be stored in the areas designated by the designer and should not exceed the permitted load. Only plant specified in the design brief should be allowed on the structure unless clearance is received from the designer.

The placing of in situ concrete should be carried out in the sequence agreed by the designer. It may be taken to the working position by crane and skip, by barrow or dumper, or be piped or taken by mixer truck. The free fall should be limited to a maximum of 1 m for specially designed falsework and to 500 mm where a standard solution has been adopted. Also, heaping of the concrete within a small area should be avoided, e.g. in an area of 1 m^2 the height of the heap above the formwork surface should not be more than three times the depth of the slab unless special allowance has been made in the design.

Checking. Immediately before loading the falsework should be examined and checked by an experienced and responsible person to ensure that it has been constructed properly and in accordance with the drawing(s) and specification (where appropriate). Only when he is satisfied and the whole of the falsework is approved should permission be given to load.

Also, throughout the loading period a careful watch should be kept to identify any structural distress and the necessary action should be taken at the first sign of failure.

Temporary works co-ordinator. British Standard 5975[4] recommends the appointment and use of a temporary works co-ordinator. Briefly it is the co-ordinator's responsibility to co-ordinate all the falsework activities such as the provision of a design brief, the production of a satisfactory design and the required drawing(s) specification and any other information, the receipt on site by the appropriate authority of the design, ensuring that checks are made throughout the construction phase and that any changes in materials or operations are transmitted back to the designer for consideration before they are used or implemented, supervising the final check and issuing formal permission to load when the checking has proved satisfactory. The co-ordinator would also be responsible for implementing the inspection during the loading stage and for advising on the action to be taken should the structure exhibit any sign of distress at this stage, and finally, would issue formal permission to strike the falsework when the permanent works had attained sufficient strength to be self-supporting.

Structural framework

The most serious accidents that occur during the erection of structural framework are due to

- erectors falling from heights when at their places of work or going to and returning from them
- people at lower levels being struck by tools or materials falling or being thrown down
- the uncovenanted collapse of the whole or part of the framework causing people to fall or striking those at lower levels.

The prevention of accidents depends to a great extent on the following.

- The designer having a good knowledge of safe erection procedures and taking these into account in the detailed design – in particular the need for temporary bracing or other measures to ensure the stability of parts (or the whole) of the framework should be clearly shown or stated on the drawings and/or in the specification; the specification may also be used to advise on any special sequence of erection necessary to ensure stability of the structure.
- Good communication between the designer and the erection contractor, for example to ensure that any measures to be used by the contractor that may put extra loading on the uncompleted structure (e.g. by the storage of material or the provision of temporary working platforms) or any proposed design modifications are discussed and approved by the designer.
- The production of and adherence to a method statement, which must be kept on site, by the erection contractor which should include detailed requirements for
  ○ the sequence of erection including aligning and levelling procedures
  ○ ground level assembly work prior to lifting
  ○ slinging and lifting
  ○ temporary bracing and support
  ○ access
  ○ safe working places and personal protection including safety nets, belts, harnesses and lanyards
  ○ the plant to be used including inspection and maintenance programmes
  ○ stacking and storing arrangements.
- Good management organisation and control and good site discipline to ensure adherence to the erection plan and the maintenance of safety features (e.g. plant, working platforms and ladders); the person in direct charge of operations on the site should be provided with a copy of the method statement and must have a thorough understanding of its contents.
- The installation of permanent stair access and the structural floor system as early as possible.

Preventing falls

Falls from the structure can be prevented by the provision of safe access to and egress from all places of work together with safe working places for positioning structural members, bolting up and other operations. However, accidents during framework erection may be minimised by reducing the need to work at heights by pre-assembling sections on the ground.

Safe access. For *vertical access* people should not be permitted to climb columns unaided or to slide down them. Vertical access and egress may be by means of ladders which may be inclined when serving single-storey heights. Inclined ladders should have wire scaffold lashings or rope lashings attached to the stiles at the upper ends and these should be fixed to a convenient beam to prevent movement. Where the lashings are taken around sharp beam corners, packing pieces should be inserted to prevent high stress points. Before columns are erected vertical ladders may be securely fixed to them; with steel I-section columns, flats (similar to rectangular section ladder rungs) may be pre-welded between flanges near the top end of a column or column length. The flats serve as the last few ladder rungs and may be used as lanyard anchor points for safety belts or harnesses when the top of the ladder serves as a short-term working place, e.g. for the erector to receive and position an incoming beam.

All ladders should be positioned to allow sufficient space at the back of the rungs for a firm footing on the ladder and for the toes of footwear to protrude. In the case of square or rectangular concrete columns timber distance pieces will be necessary between column and ladder stiles.

For *horizontal access* steel erectors should wherever possible use properly fixed lightweight staging or scaffold boards. The staging must have adequate bearing, be properly and evenly supported and secured, and the span must not be excessive for the load to be carried. The walkways should have a minimum width of 600 mm and be provided with suitable guard rails and toe boards along each side where a fall of more than 2 m is possible. The fixing and moving of horizontal access ways should be carried out from a safe position or by erectors wearing safety harnesses with lanyards securely anchored.

A commonly used method of moving across framed structures is along the beams themselves. *Beam walking* along the top surface of concrete beams or flanges of steel beams is unacceptable and should not be allowed – it is very dangerous and has led to many serious accidents.

One method of crossing is by straddling the beam. It should only be used for I-section beams. The erector sits astride the top flange with his feet (heel and toe) resting on either bottom flange. The erector then grips either side of the top flange with his hands and moves along, maintaining this sitting position. For this method the depth of beam must be suitable to allow the sitting position described and there must be no obstruction such as cross beams which necessitate a change in position. The method should only be used for access of short duration.

Where straddling is not convenient and access is of short duration, walking along the bottom flange on one side of the beam may be used. The top flange should not be higher than the erector's waist when he stands on the bottom flange, and the top flange width must be such that he can comfortably reach across and obtain a secure handhold. He is then able to move sideways along the bottom flange.

Means of access that also provide safe working places are

Fig. 4.17. Use of telescopic hydraulic work platforms and tandem lift of frame for Blackburn Rovers' grandstand (courtesy Hewden Stuart Limited)

- power-operated mobile work platforms (Fig. 4.17)
- man-riding skips
- tower scaffolds
- common scaffolds.

For power-operated mobile work platforms a firm base is required; the platform stands at ground level, normally on the already installed concrete sub-floor. Man-riding skips are suspended from a crane or winch, and controlled by a deadman's handle, i.e. the brake is applied when the control level or switch is not being held in the operating position. Power lowering should also be provided. The skip should be not less than 910 mm deep and be provided with an anti-spin device. Consideration should be given to securing workers in the skip by means of safety belts or harnesses attached by lanyards to a secure place on the skip. For tower scaffolds a firm level base is necessary – again these are generally used on the already

(a)   (b)

Fig. 4.18. (a) Temporary work platforms in use on Queen Elizabeth II Bridge, but steel erectors must still wear harnesses; (b) steel erectors wearing body harnesses and lanyards securely attached (courtesy HSE and Trafalgar House Construction)

installed concrete sub-floor or in a multi-storey building on the floor beams when these have been positioned.

Safe working places. Where a good deal of bolting or welding at height is necessary a properly constructed working platform with guard rails and toe boards may be provided, preferably fixed at the required position on the structural member before its erection (Fig. 4.18). Such working platforms have to comply with the requirements of the CHSW Regulations.

Some types of lightweight prefabricated platform with guard rails and toe boards, barriers or other protection to comply with CHSW Regulations are constructed to be lifted up and hooked over structural sections. Other methods of providing safe working places at height include

- ladders for work of short duration: where these are attached to columns a safety belt or harness should be worn and the lanyard should be attached to a secure place
- tower scaffolds, useful for single-storey work or for work from completed floors of multi-storey frames (Fig. 4.19)
- power-operated mobile work platforms which could stand on completed concrete slabs or floors, provided that the floor is of adequate strength (*see also* Fig. 6.2)

Fig. 4.19. Tower scaffold access for frame erection (courtesy HSE)

- man-riding skips, suspended from a crane or winch, which are useful (similarly to mobile platforms) for bolting-up work, purlin erection, truss positioning and sag bar erection (*see also* Fig. 4.39); this type of plant makes it unnecessary for erectors to walk up rafter backs to position purlins – a very dangerous operation
- sheeting rail cradles, consisting of two small working platforms with guard rails and toe boards separated by two supporting beams, a bay-width apart; lifting points with shackles provide four-point suspension by a chain sling for lifting and lowering by crane; a rail carrier is incorporated in the cradle, and with one man on each platform, sheeting rails can be erected for the full height of the structure bay by bay.

Safety belts and harnesses (*see also* p. 262). Where working platforms cannot reasonably be provided, personnel should wear safety belts or harnesses, attached by a lanyard to a secure place. This may be to pre-welded flats, or to the eye bolt of a girder grip device, e.g. the Manlock.

Choke-hitching of lanyards around structural sections is not recommended because of the very high stresses, sufficient to cause failure, that occur in the lanyard in the event of a fall. An alternative is a tested wire rope loop of minimum diameter 8 mm with thimble eyes at each end. The loop is passed loosely around the structural section and the two thimble eyes are threaded into a carabiner shackle attached to the lanyard. The lanyard anchorages should always be above the user to limit any fall distance. In any case, the fall height should not exceed 0·6 m for belts and 2·0 m for harnesses.

For the erection of masts and towers which involves a good deal of movement about the structural frame and where other methods of providing safe working platforms are impractical, retractable lanyards (Fig. 10.9) or taut steel wire running lines with a safety harness may be used. The structural framework and the anchorages must be sufficiently strong, however, to resist the forces generated by an erector falling.

Safety nets. In some fairly large structures consideration should be given to the use of safety nets as an added precaution to prevent or minimise injury in the event of someone falling, but they must never be used as an alternative to the proper working platforms and places described above. A decision should be made during the design stage and suitable fixing brackets should be fitted to the beams or columns before erection. Nets should be provided and used in accordance with the latest version of BS 8093: 1991 *Code of practice for the use of safety nets, containment nets and sheets on construction works.*[5] They must be slung as close as possible to the working areas but never lower than 6 m below them. Good access should also be provided to the fixing points and all measures should be taken to reduce the dangers involved in fitting the nets.

Use of cranes for erection. In addition to losing their balance and falling from a height, erectors can also be knocked down from their working place. This risk can be minimised by the careful control and handling of structural elements when they are being lifted into position. The care and expertise of the crane driver is an essential element: if it is not possible to see the load for the full extent of its travel, the driver must be guided by a banksman, and the communication between the banksman and driver must be of a good standard. The crane driver must carefully observe not only the travel of the load but also the travel of the jib to avoid colliding with the structure – which might throw an erector off balance. Tail ropes should be fitted to components and handled to eliminate unnecessary swing and the load should be moved slowly. Components to be lifted must be properly slung to prevent sections falling and should have packing pieces at the corners; moreover the crane support should not be released before the component is properly secured. The erectors themselves must be occupying a safe place either on a properly constructed platform or, where straddling a beam, standing on a beam or standing on the rungs of a ladder, they should be wearing a safety belt or harness with the lanyard effectively anchored to the structure.

Care must be exercised in slinging and transporting materials and components by crane or winch to prevent their becoming loose and falling. Particular care is needed in releasing slings and the use of remote-release shackles should be considered. Where a pole (stick) is used for lifting purposes it should be erected on an adequate foundation to avoid settlement, and should be properly guyed, with the ropes not fouling the framework.

The practice of using more than one crane to lift one component or a single assembly should be avoided wherever possible. However, sometimes *tandem or multiple lifting* has to be resorted to. In this case

- only one person, competent and experienced in tandem lifting, must control the lifting operation; this person must be in direct communication with the crane drivers
- all involved in the lift must be experienced and fully briefed about their role and the method of carrying out the lift
- where for the lifting operation the cranes are required to move while carrying the load the paths they are to follow must be clearly marked; only the person who is directing the operation may give instructions to the drivers and hence control the movement of the cranes
- the safe working load of each crane for the required jib length and operating radius should be at least 25% in excess of its calculated share of the load to compensate for load shedding from one to the other; this may be caused by uneven ground, load swinging, variation in the relative movement between each crane, or differential bounce effect as the load is moved along uneven ground.

Unless great care is taken in the planning and execution of this dangerous manoeuvre, there is a good possibility that the load will be dropped due to the overturning or overloading of one of the cranes involved (*see* Figs 4.17 and 4.25).

Preventing tools and materials falling

Wherever possible work should be programmed to avoid the need for workers to be underneath areas of higher level working. Barriers and warning notices prohibiting access should be erected at the approaches to all danger areas. Where it is not possible to exclude work activity below the erectors the people below must be provided with adequate overhead protection in the form of material nets, sheets, fans or some other form of structure to give protection from falling objects.

Erectors should be provided with suitable frogs on their belts to carry podgers, wrenches and other tools of their trade. Bolts and nuts should be kept in suitable containers provided at all working places or they may be kept in a pouch on the erector's belt. Tidiness and care with small items are essential; loose bolts, washers and tools should not be left on beams, walkways or working platforms. The use of toe boards on working platforms reduces the likelihood of items being kicked over the edge. Small items should never be thrown to the ground or from one erector to another.

Stacked materials. Accidents may arise from the collapse or fall of poorly stacked steelwork in the storage area. Lighter members should be stored on purpose-made racks of adequate stability. The height of stacked materials not stored on racks should be restricted and the materials should be supported on timber battens with packing and wedges to prevent the members slipping and the stack collapsing. Materials should be stacked clear of the ground because deposits of mud on them may cause a slipping hazard when they are in position at a higher level. The stacking should be arranged so that slings can be fixed and removed safely and safe means of access should be provided where it is necessary to climb on to a stack for selection and slinging.

Preventing framework collapses

At all stages during the erection phase most frameworks, or parts of them, are in unstable equilibrium. This may be due to

- the geometry of a series of pin-jointed frames
- the dependence for stability on the structural interlocking of other elements such as cross walls, floors, cladding panels or other elements to be erected at a later stage
- the temporary release of bolts and fastenings to facilitate aligning and levelling.

Relatively small forces on a structure due, for example, to the slight impact of a crane load or jib, eccentric loading from material stacked on the framework, overtightening of guy ropes, freeing jammed members, springing or pulling members into position, wind and even someone climbing a ladder positioned against a light framework, may be sufficient to cause collapse. Adherence to a properly prepared method statement will overcome these problems. There is also a need for the designer to convey to the erector information about temporary instability.

Framework stability. The main principle for ensuring stability during erection is to create a rigid box consisting of four columns connected at their tops by beams, the frame being prevented from distorting by bracing between the columns. Single columns must first be anchored by guys or props and wedges (in the case of precast concrete columns in foundation pockets) in four directions although when the adjacent column has been erected and connected and braced to it, the two guys in that direction may be removed.

When the rigid box has been established, the rest of the framework is

(a)

(b)

Fig. 4.20. (a) Collapse of precast concrete framework due to lack of temporary bracing; (b) reconstructed frame properly propped and guyed (courtesy HSE) built from it and effectively anchored to it. However, temporary bracing, column wedging, propping and guying will still be required for the various elements of the structure as it 'grows'. In no circumstances should sheeting rails or purlins be relied on as temporary bracing.

Positioning bolts may need to be slackened during the lining and levelling operation. To minimise the likelihood of collapse final alignment of the framework should be carried out a bay at a time as the erection proceeds (no more than two bays should be erected before final alignment is carried out) and permanent connections then made and holding-down bolts and baseplates fully grouted. During the alignment all temporary supports and bracing should be maintained. Similarly, when joints are loosened to facilitate the positioning of other members the stability of the framework must be maintained by means of temporary guys, props or bracing.

Trusses and girders. Steel roof trusses and lattice girders should be propped and/or guyed into position before the lifting gear is released. An adjacent truss or girder should then be erected and the necessary plan bracing installed to provide stability (for girders diagonal bracing may also be required between the top flange of one and the bottom adjacent beam), connecting the two before the lifting gear is released. Reliance must not be placed on purlins between trusses to provide the required stability.

A common fault in the erection and positioning of roof members which has led to many collapses concerns the installation of purlins. There is a tendency, when one or two trusses have been positioned, to lift and park the majority of the purlins in one place between two adjacent trusses. The erectors then take them from this stack and bolt them in their final positions between the erected trusses. A large number of purlins parked in a narrow area may be sufficient to cause an overload and, especially with the relatively slender rafters of a portal frame, may cause buckling and collapse. Moreover, erectors tend to drag the purlins from the stack across the backs of the rafters to their final position, and the force required to overcome the friction is very often sufficient to cause the overturning of a truss. Either the specially designed purlin cradle referred to earlier should be used or the quantity of stacked purlins should be limited and spread between a number of trusses – and, of course, they should be lifted and not dropped into position.

Precast concrete. Precast concrete columns are usually located on level packings in pockets cast in the concrete foundation. Wooden wedges cut to match the slope of the pocket sides should be inserted on all four sides before the column is released from the crane. Stability should also be provided by four push-pull props located against bolted-on timber yokes whose lower faces are chamfered to bear fully on to each prop. The foot of each prop must be firmly anchored at floor level. When two or more columns have been erected and connected by a beam or beams, temporary bracing by taut wire rope or scaffold tubes must be installed. Propping and wedging of the columns must be left in position until the foundations are fully grouted and the permanent connections between columns, beams and rafters have been made. (*see* Fig. 4.20)

With precast concrete portal frames, where the first rafter is lifted in one piece and connected to the columns the lifting gear must not be released until temporary stability is provided to the rafter by means of props or guys in two directions at right angles to the span. On large spans the rafter may be in halves, in which case both halves need temporary support before the lifting gear is released. After the erection of the adjacent rafter the permanent bracing between the two should be installed followed by the purlins. Finally, after all the permanent connections have been completed the temporary supports, bracing and associated brackets or cleats etc. may be removed.

Roofwork

Work on or from roofs is a high risk activity. According to the HSE guidance booklet *Safety in roofwork* (HS/G 33) almost one in five of all construction deaths occur in roofwork. The activity includes the construction, demolition and maintenance of roofs. It also includes repairs and maintenance etc. to chimneys, rooflights and ventilators on roofs.

Risk assessment

Before any work is carried out (even of short duration) a risk assessment should be made after which a method statement should be drawn up. This should identify step by step a safe system of work for the task to be undertaken. It should take into account the relative statutory requirements, in particular the CHSW Regulations and, where appropriate, the protection of the public and other site workers. Once the method statement has been evolved it must be drawn to the attention of all those involved particularly the workforce and those immediately responsible for them on site, usually the charge hand or foreman. A copy of the method statement should be kept readily available on site and its whereabouts known to those immediately responsible for the work.

Training and equipment

Roofwork is a construction activity where training, technical knowledge or experience is necessary to reduce risk of falls. It must only be carried out by those who have these skills or are under the direct supervision of someone who has.

Access

The general safety rule that applies to so much construction work is the basis for securing the safety of roof workers. It is to 'provide safe access to,

Fig. 4.21. Good tied ladder access to secured crawling board for roof maintenance

and egress from every place of work and to provide and maintain a safe place from which the work can be undertaken'. Safe access to the roof will generally be by ladder and the advice given on the use of ladders in the section of this chapter on scaffolding (p. 80) should be followed (Fig. 4.21).

Flat roofs

If the work is to be carried out on a flat roof or one up to 10° pitch, edge protection must be provided in the form of guard rails and toe boards or barriers described in the CHSW Regulations to prevent workers, materials, tools, plant, etc. falling from the edge. Examples of proprietary systems are illustrated in HSE Guidance HS/G 33 (*see* Appendix 4). For maintenance and repair work on flat roofs it is important to identify the construction and type of roof covering and its condition – accidents have occurred when workers have stepped on to an area of roof that has considerably deteriorated which has then collapsed under their weight.

Sloping roofs

On non-fragile sloping roofs the method of access up or along the slope to the place of work will depend upon the degree of slope, the nature of the surface and the weather conditions. On relatively low pitched roofs with a non-slippery surface providing a good foothold it may be acceptable to walk on the surface. However, if the weather conditions have made the surface slippery then suitable crawling boards must be used. They must

Fig. 4.22.   Result of the mistaken belief in the apparent safety of sheeting over roof purlins

also always be used whenever the pitch is greater than 30°. These boards must be properly supported and securely fixed to the sloping surface or over the ridge or in some other effective manner to prevent movement.

When work is to be done on or from a sloping surface it may be carried out from crawling boards provided that edge protection barriers designed and constructed to prevent a person or tools or materials falling are positioned at the eaves (Fig. 4.23). Suitable arrangements are illustrated in HSE Guidance HS G/33 (*see* Appendix 4). However, the barriers need not be provided if the fall from the eaves is less than 2 m or if the work to be carried out is of a minor nature of short duration (e.g. replacement of a tile or examination of a small area) or if the work on the roof is carried out from a properly constructed working platform in accordance with the CHSW Regulations.

When crawling boards are anchored over the ridge the 'anchor board' should bear on the opposite slope and *not* rely on the ridge capping. In such cases it is advisable to use a purposely designed and manufactured ridge iron. Eaves gutters of the ogee or half-round type should never be relied upon to support roof crawling boards. Generally, where a tiled or slated roof is being constructed or repaired and the roof battens provide a secure foothold and handhold i.e. strong enough to bear a person's weight, then crawling boards are unnecessary.

Roof sheeting

The fixing or removing of non-fragile roof sheeting is a particularly hazardous activity. The dangers of working at the leading edge may be

Fig. 4.23. Good protection at the lower edge of a sloping roof

overcome by working from suitable crawling boards (at least two side by side), incorporating guard rails and toe boards to protect workers from falling backwards. In addition a safety harness should be worn secured to an adequate anchorage point to enable the wearers to safely lay the sheeting in the gap between the edge of the crawling board work platform and the previously laid sheet, or to remove it as the case may be.

Fragile roofs

Falls through fragile roof coverings e.g. asbestos cement, especially during maintenance and demolition work is a major cause of fatal and serious injury. Sufficient and suitable crawling boards properly supported and secured to prevent movement must always be used for access, egress and as working platforms unless a properly constructed working platform is used (Fig. 4.23).

At no time must there be a need to step on to the roof surface for example to move crawling boards. It is a mistaken belief by many roofworkers and demolition workers that it is safe to walk along, or even stand on, the line of hook bolts securing the sheets to the purlins. This practice should never be used (Fig. 4.22).

To comply with the requirements of the CHSW Regulations no one must pass across or work from fragile material from which the fall would be 2 m or more. No person shall pass or work within 2 m of fragile material where such a fall is likely unless guard rails, covering or other suitable means are provided to prevent falls through the material. Also prominent

Fig. 4.24.   Safe access over sloping fragile roof

warning notices must be displayed at the approaches to places where fragile material is situated.

Valley gutters are often used as access ways to adjacent roof slopes for maintenance purposes. If the roof covering is of fragile material it is necessary to prevent anyone who may lose his balance falling through the adjacent roof. One way of obviating this is to provide suitable timber covers to the roof slope on each side of the gutter. These should extend for the length of the gutter and for at least 2 m up the adjacent slopes from the gutter. They should be stout enough to absorb the load of a person falling and constructed so as to transfer that load to the roof trusses. Alternatively a safety harness with lanyard attached to a taut, properly anchored safety line may be used.

Where their use is possible, hydraulic working platforms, scissor lifts or tower scaffolds may provide a safe alternative to working on or from fragile roof surfaces.

Cranes

The main hazards associated with crane operations on construction sites are

- the overturning of a crane or the structural failure of one of its elements
- the dropping of the suspended load or part of it
- electrocution
- trapping of people
- incorrect erection and dismantling procedures.

These hazards are considered in respect of mobile, tower and derrick cranes, which are the types most commonly used on construction sites. More detailed information may be obtained from British Standard *Code of practice for the safe use of cranes* CP 3010 : 1972.[6]

Planning lifting operations

Virtually all crane accidents are the result of operational errors rather than inadequate design or lack of quality in the manufacturing process. It is fundamental for safety, whether for one-off jobs or routine activities, that all lifting work is carefully planned. Planning should include

- selection and briefing of the lifting team, i.e. drivers, slingers, banksmen and supervisors
- organisation of the lifts: for major lifts this would need to be very detailed
- identifying the weight and the centre of gravity of the load to obtain the correct position for lifting; where either has to be calculated this must only be done by persons competent to do so
- the maximum radius to which the load will be taken
- the maximum lift height for the load
- selection of the crane(s) and lifting gear to be used, allowing an adequate working margin for load, radius and lift height, so that the crane(s) is not working at maximum capacity for long periods
- the siting of the crane(s) and the necessary ground preparation: for the siting, the path to be taken by the load and moving parts of the crane body and jib etc. throughout the lift should be considered relative to any obstruction; it should also include ease of access and suitability for erection and dismantling operations; ground preparation should include levelling and foundation requirements; the routes to be taken by mobile cranes should be assessed for their suitability; they should be clear of obstructions and not too close to the edges of excavations or embankments; the ground must be reasonably level with a bearing capacity suitable for the wheel or track loads – sleepers or mats may be required
- special considerations for any tandem or multiple lifting (*see* p. 95)
- erection and dismantling procedures in strict accordance with the

methods, sequence and materials laid down in the manufacturer's manual

- programming operation to avoid working under suspended loads; erection of notices warning people not to pass under suspended loads.

Selection of operatives

Drivers. All crane drivers must be at least 18 years of age, medically fit and with an aptitude for judging distance. They should have been trained in crane operation, simple maintenance and hand signalling. It is advisable that they hold a certificate of competence for the particular machine they operate. The Construction Industry Training Centre holds courses and issues certificates to those who successfully complete them (*see* p. 246).

Banksmen and slingers. Banksmen and slingers must also be at least 18 years of age, medically fit and with an aptitude for judging distance. They should have been trained to a suitable level of competence. They should be readily identifiable, for example they should wear a garment that is easily visible or a distinctively coloured safety helmet.

Duties of the lifting crew. Where the driver cannot see the whole area of the lifting operation, from picking up the load, throughout its full travel to where it is to be put down, a banksman must be employed to guide the driver and to control the operation of moving the load. Communications between driver and banksman must be of the highest order and are best made by radio telephone. If visual hand signals are used both driver and banksman should be clear about their meaning. They should preferably conform to those recommended in British Standard *Code of practice* 3010.[6] The distance between the driver and the banksman should be such that hand signals are clearly visible and there is no possibility of confusion of their intention occurring. Where this is not possible, and radio telephones are not available, more than one banksman must be used. One of the duties of the banksman is to prevent people from straying underneath the carried load. One person particularly vulnerable is the slinger. When the load is ready for lifting the slinger should be positioned out of the line of lift and travel before signalling to the banksman or driver to start lifting, and should never walk or stand under the suspended load.

Main hazards and safety measures for crane operations

Overturning or structural failure. Overturning or structural failure in cranes is caused by overloading. It is essential that the safe working load at its associated radius is never exceeded. Sometimes the manufacturer's safe working load figures include the weight of the lifting gear (hook, slings, spreader beams, etc.) and therefore the weight of these items must be deducted from the manufacturer's figures to arrive at the load that may be lifted.

The safe working load may be determined by the stability of the crane, i.e. when the safe working load is exceeded by a relatively small amount, tipping will begin. However, for some cranes the safe working load may be limited by the structural or mechanical strength of components such as the jib or hoist ropes. A structural failure will then occur without any indication of tipping – this is often the case for mobile cranes operating with outriggers in position and for hydraulic telescoping cantilevered-jib truck cranes. It is essential therefore, that drivers and planners do not rely on loading a crane to the point where it shows early signs of tipping in order to obtain the highest possible lifting capacity. Structural failure may occur first, suddenly and without warning.

Overloading is not necessarily a load in excess of the load specified as the safe working load but may occur because the allowable radius for the load is being exceeded. Some examples are

- excessive load swing due to rapid deceleration of the slewing motion by severe application of the brake; similarly, snatching the load from rest and hard braking to arrest a descending load can also cause excessive swinging
- using the crane to drag a load, usually from outside the radius associated with the safe working load for that load; dragging may cause a considerable increase in the load being moved because of the frictional force incurred and should never be allowed
- eccentric lifting: the hoist rope must be vertical above the centre of gravity of the load otherwise the load will tend to swing
- a wind effect on loads of relatively large surface area (e.g. shutters): it is necessary to counter this by means of tail rope(s) attached to the load and manipulated by a banksman
- differential settlement of the crane supports or foundations: with mobile cranes where the load of the crane transmitted through the wheels, tracks or outrigger feet exceeds the allowable ground bearing pressure, sinking will occur; this will almost certainly cause a swing of the jib and load; sleepers, mats, plates under outrigger feet or wheels or even the provision of a properly constructed concrete hard standing may be used to ensure that the allowable bearing pressure is not exceeded; the maximum load to be distributed must be used in any calculations, i.e. the load under a wheel, track or outrigger when the jib is operating over them; care must be exercised to ensure that the crane wheels or tracks do not approach too closely to the edges of excavations or embankments
- stretch in the jib suspension ropes: with mobile and guyed derricks when the load lifted is near the maximum safe working load any stretch in the jib suspension rope will allow the jib head to deflect, causing the radius to become excessive; before such a lift is made, the jib should be derricked in sufficiently to compensate for this

- rebound of the jib head when the load is grounded: when the jib is near its minimum operating radius it may swing backwards when the load is grounded because of the sudden release of tension in the jib guys; this may result in structural damage to the jib by the backstays or in the crane overturning backwards; gentle and smooth grounding of the load should overcome this hazard.

Dropping the suspended load. Correct slinging of the load and its security on the hook is of primary importance. Only slings and lifting gear of the appropriate safe working load with the required valid test certificate (i.e. which have been thoroughly examined within the previous six months and appropriately certified) may be used. Chains must not be knotted and shackles must be fitted with proper pins. Slings should be protected around sharp corners by fitting packing pieces. Hooks should be of sufficient capacity for the heaviest load intended to be lifted.

The load should be slung so that the hoist rope is vertically above the centre of gravity of the load to prevent tilting or swinging. Where there is some doubt about the position of the centre of gravity, the load should be trial lifted just clear of the ground. If it tilts or swings it should be lowered and the slinging readjusted (Fig. 4.25).

Fig. 4.25. A tandem lift of an irregular load. Ensure that the load is capable of being safely lifted. Check centre gravity to prevent swing and to equalise lifts. Provide ample packing to prevent damage (courtesy Hewden Stuart Limited)

The hook should be either of the Liverpool or C-type or be provided with a safety catch to prevent the load, sling, ring or other suspension device from being displaced. The hoist rope must be of the type and size specified in the crane manual, in good mechanical condition and properly maintained. Hoist rope failure resulting from poor maintenance has been the cause of many accidents. Broken wires and corrosion are usually the first signs of a dangerous condition in a hoist rope.

The load must be carefully and smoothly manoeuvred throughout its travel and it must not be allowed to collide either with any part of the crane or with any temporary or permanent works, plant or stacked materials. The load must be maintained vertically under the hoist rope and it may be necessary for the banksman to control the load by a tail rope to prevent swinging.

Danger of electrocution. If the jib touches or closely approaches a live overhead power line the electric charge will pass into the crane body via the jib and everyone who is in the vicinity, especially the banksman if he is steadying the load by means of a tail rope, is in danger of being electrocuted. All overhead lines must be treated as live unless the electricity authority has declared them dead.

When work is to be carried out in the vicinity of overhead lines the crane should be prohibited from approaching nearer than 6 m plus the total length of the jib and including any intended extensions measured horizontally along the ground from vertically below the nearest conductor. Allowance should be made for the lateral movement of the conductors because of the effect of wind – which is very likely to take place with overhead lines of fairly long spans. The limits should be marked on the ground by warning markers parallel to the conductors.

When cranes are to pass under overhead power lines, the allowed crossing route must be clearly marked and goal posts (Fig. 4.26) should be erected at each side of the crossing approach. They should be positioned no closer to the outermost overhead conductor on each side than the horizontal distance markers. The cross-bar should be at a safe height so that the jib must be lowered below it before making the crossing. Devices are available which can be fitted to crane jibs to give a warning when they approach within a predetermined distance of an overhead conductor. However, they have limitations, and their availability should not preclude the use of horizontal markers and goal posts. When work is to be carried out within 15 m of overhead lines suspended from steel towers or 9 m from those suspended from concrete, timber or steel poles, the electricity company should be advised well in advance.

Clear instruction on the actions to be taken if the crane becomes live must be given to every crane driver and a notice giving the information should be prominently displayed in the cab. These instructions should inform the driver what to do if the crane makes contact or becomes live. The driver should

Fig. 4.26. Goal posts to indicate maximum height for the safe passage of cranes and other plant under high-tension cables

- remain in the cab
- warn others to keep away and not to touch any part of the crane, rope or load
- ask someone to inform the electricity authority
- try to move the crane away from the power line
- where this is not possible, stay in the cab until safe conditions are confirmed
- where it is essential that the driver leave the cab, he should jump clear and not climb down; no part of the crane or load and the ground should be touched at the same time.

Trapping. During slewing operations, many serious accidents have occurred with people being trapped between the moving superstructure of a mobile crane and any nearby fixed object such as adjacent structures, plant or stored materials. A clear passageway of 600 mm must be maintained between any part of the crane liable to move and any fixed object, otherwise the gap must be effectively fenced off.

Slingers are the likely victims of the other main trapping hazard, with injuries to fingers, hands and feet when they become trapped between the hook and the sling or other lifting gear as the crane takes the load. The driver must not operate any crane motion until the slinger has given the appropriate signal from a safe place. Moreover, the slinger must not grasp the hoist rope (e.g. to guide the load) at the instant of lift or just before grounding since trapping between the rope and hook block sheave is likely. A pole or tail rope should preferably be used to guide the load but if the hand is used it should be the flat of the hand against the load. When guiding the load, just before final grounding the slinger's feet must be well away from the load.

Erection and dismantling. Many accidents happen during the erection and dismantling of cranes, including erectors falling, being struck by falling tools or materials, and being trapped. Only erectors trained and experienced in the particular make and model of crane should be used for this work and they must be under the immediate supervision of a competent supervisor experienced in all aspects of the operation. The work must be carried out strictly in accordance with the manufacturer's procedures manual. Where the jibs can be extended by the introduction of intermediate sections, the jib should be properly supported on each side of the joint being disconnected before any connecting pins are knocked out, to prevent the jib falling. Also, the inserted sections must be put in the correct way round and in the correct place in the jib assembly.

All erectors should wear safety helmets and, especially in the case of tower and derrick cranes, safety harnesses and lanyards should be worn and used when working at height. With tower cranes, at each stage of the erection, the necessary counterweights or kentledge must be secured in the appropriate position. Where the counterbalance is made up of individual weights on the counter jib, these must be bolted together or otherwise prevented from becoming displaced. All pieces of kentledge and counterbalance should have their weight clearly marked on them. Before any section is lifted from the horizontal to the vertical, a check should be made for any loose tools or materials.

Safety measures for specific types of crane

Some additional points on overturning and structural failure hazards and safety measures associated with specific types of crane are now dealt with.

Mobile cranes

Operating on a slope. When mobile cranes are travelling, or working in a static position on a gradient, where the load is on the downhill side the operating radius is increased. Where static work is carried out across a slope the load will swing sideways. This circumstance should be avoided. Static work should only be carried out on level ground or when the crane has been properly levelled.

The slewing ring of the crane must always be level before any lifting operations are carried out. This must be checked at regular intervals.

Fig. 4.27. Mobile crane with outriggers fully extended and well supported on timber baulks and steel plates (courtesy Hewden Stuart Limited)

(a)

(b)

Fig. 4.28   (a) *Never* put an outrigger direct onto a concrete floor without first checking the need for timber spreaders. (b) *Never* trust a concrete or stone surface (courtesy Hewden Stuart Limited)

Where there is a need to travel on a gradient either the load should be carried on the uphill side (for travel either up or down the slope) or, where the load has to be carried on the downhill side, the operating radius should be reduced to compensate for the slope, and in both circumstances the slewing brake should be engaged. In either case the driving should be slow and steady commensurate with the ground surface.

Operating with outriggers not fully extended. Where outriggers are fitted, where they are not fully extended (they should be pinned in position), the safe working load associated with the extended outriggers is not available and, if it is used, an overload is inevitable. Outriggers must always be fully extended and jacked down on to suitable plates and packings in order to distribute the load (Figs 4.27 and 4.28).

Operating with under-inflated tyres. Where the tyres of a wheeled mobile crane are not properly and evenly inflated the crane will not be operating on a level base and an excessive radius will occur.

Assessing the strength of jib and suspension ropes in piling operations. A particularly onerous duty for mobile cranes on construction sites is their use in piling operations. The crane must be of adequate capacity safely to accommodate the weight of the piling equipment (e.g. hammer, extractor and leaders). When the crane is moving from one location to the next with leaders or frame suspended from the jib, the suspended equipment must be prevented from swinging and subjecting the jib to undue lateral and dynamic (bounce) loading.

When sheet piles or the casings of bored piles are being extracted an extractor should be used to overcome the friction between the ground and the pile or casing. Final extraction should be carried out by a smooth steady pull on the hoist rope. This is a very heavy duty which should be restricted to machines that are designed for it. When the required capacity of the crane is being assessed for extracting sheet piles, the friction between adjacent piles should be allowed for. It is of paramount importance with this type of work that the pull is exerted on a vertical hoist rope.

Tower cranes

Wind effects. Tower cranes are specially vulnerable to the effects of wind and must not be operated outside the limits stipulated by the manufacturer. An anemometer should be fitted at the highest suitable position with its indicator in the cab clearly visible to the driver. When lifting operations are suspended for any reason, all loads should be taken off the hook and the hook should be raised to its highest position. The main jib should be positioned away from the wind and put into free slew. The power supplies to all motions should be switched off and the hoisting and trolleying (but not the slewing) brakes or locks applied. In adverse wind conditions the wheels of rail-mounted cranes should be clamped or movement prevented by other means. No advertisement boards should be attached to any part of a tower crane without the manufacturer's approval and then only in strict accordance with the restrictions imposed.

Collision between tower cranes. When more than one tower crane is to be used on a site, where they are in close proximity to each other there

Fig. 4.29 Audio-visual warning devices are often fitted to tower cranes to warn drivers of possible collision of jibs, ropes, counterweights or loads (courtesy Hewden Stuart Limited)

is a danger of jib collision, jib-hoist rope collision and counter jib-hoist rope collision. Careful planning and skillful and well-controlled crane operations are necessary if serious accidents from overturning, structural failure or load displacement are to be avoided (Fig. 4.29).

Where possible, cranes should be sited so that jibs, hoist ropes and counter jibs cannot touch. Where this is not possible, even when the jibs are at different heights, careful control of all lifting operations is necessary with strict adherence to the programmed lifting sequences. These must be agreed at the planning stages and controlled by one person. Good direct communication between all drivers and the ground controller is essential and the maintenance of communications equipment is vital.

Devices are available to give warning when the crane is approaching an obstruction but these should only be used as additional safeguards. Where more than one rail-mounted crane is operated on the same track, cut-out switches should be installed on the track to prevent collision.

Foundations for tower cranes. Foundations for tower cranes must be properly designed to CP 3010: 1972[6] and in accordance with the manufacturer's recommendations. It is particularly important that the allowable bearing pressure of the ground is not exceeded and that the bases are well drained and maintained during operations.

For rail-mounted cranes, where the rails are supported by sleepers, these must be spaced to support the maximum crane wheel loading and to resist any tendency to heel or turn the sleepers over. Sleepers should be at right angles to the tracks and should extend sufficiently at both sides of each rail to ensure that the load is properly spread to the foundation. Where a layer of hardcore and ballast is used beneath sleepers, it should be maintained to provide a solid and compact support. It should not be packed high in the centre of the sleepers so as to cause convex bending.

The tracks should be laid straight or to radii of the correct curve to suit the types of bogie, and they should be tightly butted end to end. They should be in good condition. The gauge should be maintained by means of tie rods which should be able to withstand the compressive as well as the tensile forces. This can be done by threading the tie rod through scaffold tubes which have been cut to the correct length between the inside webs of the rails.

Fish plates should be fitted to both sides of the rails and secured by bolts. The holes in the rails to accommodate the fish plates should be drilled, not burned, and located over sleepers. Security from running off the end of the track should be provided first by travel limit switches placed near the end of the track, then by sandboxes to reduce the travelling inertia of the crane if it overruns the limit switches. Finally, at a distance equal to half the crane base length from each end, crane stops should be securely positioned.

Climbing cranes are often supported by the permanent structure and it is important that the erectors check with the client or their professional advisers to ensure that the structure is capable of withstanding the expected loads.

Danger of electrocution. Where the crane is electrically operated from a source external to the crane, the structure and all electrical equipment must be properly earthed. For rail-mounted cranes at least one of the rails must be electrically bonded across every joint and the rail itself properly earthed.

Power supply cables should be of armoured construction or otherwise protected from mechanical damage (e.g. run in conduit), and where the protection is provided by a conducting material it should be bonded to earth at each end.

Access for personnel. Good access to reduce the possibility of falling should be provided for all those who need to climb the crane (e.g. drivers, maintenance workers and inspectors). A permanent steel ladder is required, securely fixed and with safety hoops, and extending from the ground to the driving cab. Wherever possible the ladder should be within the mast, and must include landings at 9 m intervals for people to rest safely.

Access through the slewing ring presents a trapping hazard between the fixed and slewing parts of the mast. Devices should be fitted to enable the slewing brake to be controlled from below the trapping area. Alternatively, access may be provided from just below the slewing ring by means of a ladder outside the mast. This presents a falling hazard when people are climbing from the ladder inside to the access platform outside the mast, and strong wire mesh barriers between guard rails and toe boards should be provided on the platform.

Safe means of access is necessary along the jib and counter jib. Any gangway or platform at cathead level, together with walkways (usually open mesh) along the jibs, should be provided with guard rails and toe boards or barriers to comply with the CHSW Regulations. Alternative access may be provided by a lightweight cage that is suspended from the jibs and can be traversed along them. In such a case it is safer if movement control is operated from the cage itself. However, it will then be necessary to incorporate interlocks so that the saddle traverse control in the cab is locked off when the jib cage is in use.

Where neither method is practicable everyone on the jibs must wear safety harnesses fitted with two lanyards, so that when negotiating obstacles the wearer can be secured at all times either to the structure or to a specially fixed taut steel-wire rope of adequate strength to withstand the shock of someone falling.

Scotch derricks

Incorrect position of the jib. The angle between the backstays of a scotch derrick is approximately 90° and no attempt must be made to erect the jib and to lift a load within this quadrant. This action might result in buckling of the backstays, uplift on the kingpost and overturning.

Kentledge. When the scotch derrick is rail mounted on bogies, the correct amount of kentledge should be secured to the backstay bogies to prevent tipping under maximum load, and the total weight of the kentledge should be clearly marked on each bogie. Also, each individual block of kentledge should be marked with its weight.

Bracing of gabbards. Additionally, when the scotch derrick is mounted on gabbards, either fixed or rail mounted on bogies, the gabbards must be adequately braced within themselves and between each other (except across the bowsill side) to overcome buckling or twisting.

Derricking under control. For a single-motor scotch derrick an interlock must be fitted between the derricking clutch and the pawl sustaining the derricking drum so that neither can be disengaged unless the other is fully engaged, otherwise the jib with its load can fall.

Guyed derricks

Some additional points concerning overturning and structural failure hazards and appropriate safety measures with guyed derricks include the following.

Unless the guy ropes are of the correct size, and are correctly spaced and properly anchored, the mast may topple. The security of the guy ropes should be ensured by means of anchor stirrups or plates set in a concrete foundation. The guy ropes should be equally spaced radially around the mast and a minimum of five should be used. The vertical angle between the mast and guy should be not less than 45°. The ropes, which are usually fitted with turnbuckles, must not be over-tensioned – all tensioning should be carried out by a competent person experienced in undertaking the functions.

A chart showing safe working loads at different radii for various jib lengths should be attached to the mast. This should also indicate the rope sizes, their angle of inclination and the anchor weights for each guy.

Testing and examination of cranes

In Britain it is a statutory requirement that all cranes on construction sites shall

- be tested and thoroughly examined before being taken into use, every four years and before use after any substantial alteration and repair
- be inspected weekly
- be thoroughly examined every 14 months
- have the anchorage or ballasting arrangements
  - examined on each occasion before erection
  - examined after adverse weather conditions that are likely to have affected their stability
  - tested after each erection on site and after any adjustments that involve changes to the ballasting or anchorage arrangements or their security
- when an automatic safe load indicator is required, have it
  - tested before the crane is taken into use and on each occasion after it has been wholly or partially dismantled
  - tested after each erection or any alteration of the crane likely to have affected the proper operation of the indicator
  - inspected weekly.

All these tests, inspections and examinations must be carried out by competent and experienced staff, and the results must be entered in the appropriate register for the first three requirements and on Form 91 for the last two.

Other safety measures

Automatic safe load indicators. Automatic safe load indicators are safety devices which provide a visual warning to the driver when the crane is lifting between 90% and 97·5% of its maximum safe working load. An audible warning is given to the driver and is sufficiently loud to be heard by those in the vicinity of the crane when the crane is lifting between 102·5% and 110% of the safe working load, i.e. when it is overloaded.

In Britain it is a statutory requirement to fit these devices to all cranes used on construction sites which have a lifting capacity over 1 t. It is very important that they should be maintained in good working order and that they should never be immobilised or tampered with so that they do not function correctly when the crane is in use.

Fig. 4.30   Collapse of tower crane due to failure of bolts securing it to the supporting grillage (courtesy HSE)

Load radius indicators. Load radius indicators are indicators which show the radius of the jib, trolley or crab of a crane at all times and the safe working load corresponding to that radius. It is a statutory requirement in Britain to fit them to all cranes which have a variable operating radius.

Transport, mobile plant and roadworks

Transport and mobile plant

Of all the construction industry fatalities for 1981–85, 21% involved transport and mobile plant.[7] The deaths included those for loading and unloading operations and people falling from plant and vehicles. The types of vehicle involved in the accidents included excavators and shovels, earthmoving equipment (i.e. crawler tractors and bulldozers, scrapers and graders), dumpers and dump trucks, lorries, vans, cars, buses, motor cycles, mobile cranes, rough terrain forklift trucks, locomotives and rail waggons, road rollers and mobile working platforms.

Well over half of site transport accidents involve people being run over, struck or crushed by vehicles moving forward or reversing, collisions between vehicles and collisions of vehicles with fixed objects (such as falsework and scaffolding). This category includes those accidents where vehicles and plant overturn by falling into excavations and down embankments. Other accidents occur when people fall from vehicles not designed to carry passengers and when equipment or material falls from vehicles when they are moving or stationary.

The hazards of working with transport and mobile plant on site are numerous but the fundamental cause of most accidents is the lack of adequate planning and the failure of management to organise safe systems of working and to instruct, train and properly supervise personnel.

The requirements of the CHSW Regulations in relation to traffic routes, doors and gates, vehicles and plant and equipment have to be complied with in the context of these paragraphs.

Site transport planning. Hazards with site transport and mobile plant can be reduced by

- the segregation of vehicles and pedestrians by barriers and the segregation of vehicles from fixed objects, including scaffolding and falsework, by the use of barriers and/or fenders and effective signing; heavy earthmoving equipment should also be segregated from lorries and light traffic whenever possible
- the provision of clearly signed and indicated routes around and into the site and safe parking areas; site roads and haul routes should be of adequate width to allow approaching vehicles to pass safely; sharp bends should be avoided or the road should be locally widened; haul roads crossing public roads should be controlled by traffic lights or manually operated stop-go signs
- effective procedures for dealing with reversing, e.g.
  - use of banksmen
  - provision of adequate turning spaces
  - organising one-way systems for loading and unloading to eliminate the need for reversing
  - fitting audible warnings when reverse gear is engaged
  - excluding pedestrians, except the banksmen, from areas where vehicles have to reverse
  - fitting audible proximity warnings or closed-circuit television in vehicle cabs
  - providing reversing stops for vehicles tipping into excavations (Fig. 4.31)
- good effective lighting of traffic routes and work areas and, where there are obstructions along the route such as scaffolding and falsework, plant buildings and materials
- regular maintenance of site road surfaces by grading, repair of pot holes etc. and dust control by regular spraying with water
- preventing vehicles from approaching close to the edges of excavations by effective barriers and good road marking; where vehicles are required to approach excavations or embankment edges to

Fig. 4.31. Wheel stops for vehicles tipping into excavations (courtesy George Wimpey & Co Ltd and CITB)

deliver materials, for example, the reversing and tipping operation should be controlled by a banksman and effective wheel stops should be used
- the erection of goal-post height gauges at approaches to overhead obstructions, e.g. electricity lines (*see* Fig. 4.26)
- clear indication of the routes of underground cables, wires and pipes.

Other measures that must be taken by management to provide for the safety of personnel include:

- the provision of a scheme for assessing the competence of drivers and ensuring that only those employees who have been authorised to do so are allowed to drive vehicles or to operate plant; nobody under the age of 18 may drive vehicles on site
- the imposition and strict enforcement of speed limits and good driving behaviour
- the provision of loading controls for vehicles to ensure they are not incorrectly or overloaded; the use of loading height gauges where appropriate
- imposing a ruling that no loading or unloading of lorries takes place while the driver is in the cab and the banksman and/or plant operator prevents people from straying under any load being slewed during loading or unloading

- the use of construction plant fitted with roll-over protection structures (ROPS) for operations where roll-over is a possibility (see ROPS below)
- the use of vehicles and plant fitted with falling object protection structures (FOPS) wherever drivers may be struck by falling material, e.g. on demolition sites (*see* FOPS below)
- training for all site employees covering the hazards and precautions of working with site transport, followed by effective supervision
- the provision and use of high-visibility garments for all those working in the vicinity of site transport, especially banksmen
- the provision of a planned inspection and maintenance scheme for all vehicles and plant, with properly kept records
- ensuring that all subcontractors, suppliers and the self-employed meet the same standards of training and supervision as contractors' employees and that they comply with all the rules laid down in relation to the control of transport on site.

The special hazards and precautions to be taken with rail transport on site are described under the section on tunnelling (*see* p. 125).

ROPS and FOPS. To implement two EC directives the following regulations for roll-over protection structures (ROPS) and falling object protection structures (FOPS) for construction plant are applicable in Great Britain.

- The Roll-over Protective Structures for Construction Plant (EC Requirements) Regulations 1988, SI 1988, No. 363. The regulations prohibit the marketing of construction plant covered by ISO Standard 3471/1 1986 (identical to BS 5527: Part 1: 1987) manufactured after May 1990 unless it is fitted with ROPS which complies with the regulations.
- The Falling Object Protective Structures for Construction Plant (EC Requirements) Regulations 1988, SI 1988, No. 362. The regulations prohibit the marketing of plant covered by ISO Standard 3449:1984 (identical to BS 6912: Part 6: 1992 *Specification for falling object protective structures on earth moving machinery: Laboratory tests and performance requirements*[9]) manufactured after June 1990 unless it is fitted with FOPS complying with the Regulations.

Market means sale, lease, hire or hire purchase. The regulations require that for each ROPS or FOPS there must be an EC-type examination certificate and a certificate of conformity which must be displayed on the equipment. Those convicted of contravening the regulations, whether a body corporate or a senior person acting for it, or both, may be fined up to £2000.

Although ISO 3471/1 1986 includes off-highway dump trucks, it is not intended that the regulations for fitting ROPS (or FOPS) apply to small-site dumpers. Even though the regulations only require facilities on the plant for the fitting of FOPS, when the plant is used in places where falling objects or material may be expected, e.g. demolition sites, FOPS must actually be fitted.

Roadworks

An aspect of civil engineering closely affecting, and affected by, members of the public is highway work. Many users of the public highway are killed or injured in traffic accidents at roadworks (*see* Fig. 6.1) and the safety of roadworks personnel is also of great concern. The number of accidents may be reduced by adequate traffic signing and by rigid adherence to a number of simple rules and recommendations.

Safety of personnel. The principal risk to health and safety of personnel working on a trafficked road arises because of the approach speed of the vehicles and the relative speed of human reaction. Sites must be managed to reduce these risks to the practicable minimum. Although motorways and all purpose trunk roads have a lower accident rate than minor roads, because of their high speed potential they can present a greater risk when obstructed by roadworks unless the travelling public is given adequate advance warning. However, the risks to personnel working on minor roads must not be ignored because many minor roads do not meet visibility and design standards and the available carriageway width is often limited.

It is particularly important in highway working that employees understand their duties under the HSW Act to preserve their own safety and that of their colleagues and members of the public. Personnel must not expose themselves to risk. Employers must consider carefully all matters of access to avoid the need for employees to walk along or across a live carriageway as far as is possible and employees must be instructed, trained and supervised to see that the safety plan is understood and complied with.

The County Surveyors' Society and the Department of Transport have published guidance for personnel working on highways, *Notes for guidance on safety at roadworks.*[10] Below is a summary of their recommendations.

- All personnel must wear high-visibility garments with retro-reflective markings as recommended in chapter 8 of the *Traffic signs manual.*[11] Weatherproof clothing should be worn as appropriate.
- Inexperienced personnel must be accompanied by experienced trained operatives.
- Only trained operatives who are fit and alert may set out signs and cones.
- Drivers and plant operators new to highway conditions must be closely supervised until properly trained.
- Personnel should not cross a trafficked carriageway unless it is essential and safe to do so.

- When placing cones from a moving vehicle, operatives must avoid leaning out into a trafficked lane.
- All visitors to roadworks sites must wear suitable safety clothing and be fully aware of safety procedures.

In regard to care of the public

- plant, materials and personnel must not obstruct or intrude into areas marked for vehicles or pedestrians. Slewing vehicles, cranes and excavators must be carefully sited to ensure that no part of the machine may interfere or collide with passing traffic
- plant or materials should not be left unnecessarily on the hard shoulder, central areas or other roadside areas where they could be a hazard to vehicles which, for whatever reason, fail to keep to the carriageway
- the problems of the blind or the partially sighted and of other disabled people must be carefully considered when defining the areas reserved for the public.

In regard to prolonged working hours

- fatigue can be induced by monotony, stress and prolonged working hours. In the interests of safety, over-tiredness should be avoided. Management must consider the risk to safety brought about by prolonged working hours having regard to the type of work, the degree of stress and the duration of long shifts.

All staff and operatives engaged on roadworks should wear safety helmets not only for head protection but also because their conspicuous colour makes the wearer more readily visible to vehicle drivers and plant operators.

Traffic safety at roadworks. Traffic safety measures at roadworks are covered by chapter 8 of the *Traffic signs manual* [11] prepared by the Department of Transport, the Scottish Development Department and the Welsh Office and published by HMSO.

Authorities, contractors and others have a civil law liability to give warnings of obstructions on the highway caused in connection with roadworks. They also have to remove redundant signs. Notwithstanding any civil law liability, failure to erect signs etc. could constitute an offence under the Highways Act 1959 and possibly also under the HSW Act. The type and layout of traffic signs and road markings for roadworks are described in topic 3 of chapter 8 of the *Traffic signs manual*. If it is necessary to hold down temporary signs, sandbags should be used. The practice of holding down temporary signs with kerbstones or other heavy objects can be very dangerous if a collision occurs.

The boundaries of all roadworks must be clearly delineated to indicate to drivers the limits of the carriageway and to protect the workforce and the works. This should be done by traffic cones and barriers and, after dark,

Fig. 4.32. Temporary vertical concrete barriers (TVCB) to protect the workforce on live highway projects as an alternative to lane closure by coning. A client's contribution to safety.

lamps must be used as well. The safety clearances for the protection of personnel must be observed. All this is described in topics 2 and 3 of chapter 8 of the *Traffic signs manual*. The effective use of temporary vertical concrete safety barriers is shown in Fig. 4.32. Barriers to protect pedestrians on the highway must be fixed, secure and immovable so that, for example, a visually disabled person is prevented from passing through.

The movement of site vehicles and their access into the site is also covered in topic 2 of chapter 8 of the *Traffic signs manual*. Wherever possible, vehicles should not stop in any part of the carriageway in use but where it is absolutely necessary, the vehicle must be within a properly signed and delineated area. When gaining access to works areas, any cones that need to be moved must be immediately replaced. Works vehicles must be conspicuous and have one or more rotating amber warning beacons as described in the *Traffic signs manual*. When stationary, all vehicles must be 1·2 m or more from the edge of any live carriageway otherwise it is necessary to provide a coned-off taper.

Traffic signs and safety measures for minor works on minor roads are described in *Safety at street works and road works*, a code of practice issued by the Secretaries of State for Transport, Scotland and Wales under Sections 65 and 124 of the New Roads and Street Works Act 1991.[12]

Underground services. The proliferation of underground cables, pipes

Fig. 4.33. Traffic safety measures and shaft works protection outside the Institution of Civil Engineers for London Underground's Jubilee Line extension (courtesy Balfour Beatty Amec Joint Venture)

and ducts on highway works is a particular concern. Contract documents usually include the special requirements of any public utilities present, describing the hazards and explaining to contractors the rules for working near live apparatus (*see* p. 198). Reference should be made to HSE guidance note HS (G) 47 *Avoiding danger from underground services* (*see* Appendix 4).

Apart from the dangers with electricity cables and gas pipes there is the risk of flooding where water mains are severed, and of pollution where sewers are damaged.

The first essential measure towards making the site safe when excavations have to be carried out is to contact the public utilities and the local authority to advise them of the work to be undertaken and to obtain as much information as possible. Contact should be made at least one month in advance of any site work. Thereafter, it is necessary to keep in close liaison with their designated representatives. When location of the services is permitted by the authority concerned, the initial excavation must be carried out by hand and with great care.

Tunnelling

A high standard of good housekeeping (i.e. site tidiness) and site discipline (i.e. the observance of laid down operational safety procedures) is essential to eliminate or minimise the hazards implicit in all types of tunnelling operation. Compliance with procedures is particularly important in tunnelling because of the nature of the work, which is carried out in a confined area. Before tunnel construction starts a hazard analysis study should be made and the working procedures planned to overcome as far as possible the problems that are foreseen. The work plan should include both evacuation procedures and rescue operations for accidents and emergencies, e.g. a fire. Contingency planning should include discussions with the local fire brigade and ambulance service, and arrangements should be made for them to visit the site at regular intervals to familiarise themselves with access to the site and tunnel face. It is also useful to rehearse the proposed rescue arrangements for likely emergencies. It is a very skilled job and must only be carried out with the necessary expertise.

Where tunnels are being driven in air at pressures above atmospheric pressure (compressed air working) the fire brigade or ambulance personnel or both may not agree to enter the workings even in an emergency. It is therefore necessary for the contractor to provide a trained fire-fighting and rescue team, properly equipped and readily available at all times while anyone is in the tunnel.

Hazards of tunnelling

Burial from ground collapse. The general principle for avoiding ground failure is to keep the area unsupported to a minimum consistent with its type and self-supporting quality. In headings and small hand-driven tunnels up to about 3·5 m in diameter the roof, sides and face may be supported by conventional timbering, as described in BS 6164 *Safety in tunnelling in the construction industry.*[13] This must be carried out only by those with the necessary experience and expertise.

For headings timbering provides the support throughout their life (i.e. until they are backfilled) but for small tunnels a permanent lining of cast iron or reinforced concrete rings is erected immediately behind the face within the safety of the timbering. The face boards are removed as required, with excavation carried out progressively downwards from the top. In poor ground the headboards are driven forward from the previous frame to provide roof support before any excavation. The timber must be of good quality and examined before use to ensure that it is sound and free from faults.

For small headings care is necessary to ensure that completed timbering is not damaged by, for example, skips or barrows used to convey spoil from the face. Tracks should be laid and maintained in line to allow sufficient clearance between the bogies and the timbering. The timber support itself must also be inspected at least once a day and thoroughly examined at least once in seven days to be sure that it is performing its function and has not suffered damage.

Larger tunnels are constructed with shields to provide the initial protection from ground failure in the roof and at the sides. The shields comprise a stiffened steel cylinder incorporating a front cutting edge. Partial excavation of the upper part of the face is carried out after removal of the necessary face boards – the remaining boards protect the operatives from failure of the face. The shield is then pushed forward by means of hydraulic jacks approximately the width of a lining segment. The shield then protects the operatives from roof failure, and the remainder of the face is excavated in safety. After the excavation phase, the face must immediately be resupported by timbering and at the same time the next ring of segments can be erected within the protection of the back part of the shield.

In better ground the face may not need timbering and excavation may be carried out with mechanical excavators or boom-mounted cutters (roadheaders) but again advancing only one ring width at a time. Full-face tunnel-boring machines incorporating a revolving cutter head within the shield may be used in soft rock. The cutters which effect the excavation are mounted on radial arms and the spoil is directed by scoops on to a conveyor belt which carries it through to the rear of the machine. The segmental lining is again erected within the safety of the shield. Hard rock tunnels may be excavated by boom-mounted cutters, by full-face tunnel-boring machines or by means of explosives. Where support is required it may be provided by the use of steel arch ribs at intervals supporting timber or steel poling boards wedged hard to the rock face.

When explosives are used loose rock may be left in the roof or sides which could fall and cause serious injury. Therefore as soon as the area has been ventilated sufficiently to enable operatives to enter, the exposed roof and sides require careful examination, and where necessary loose rock pared down before the support work is carried out. Other methods of making doubtful areas of hard rock tunnels safe are by means of rock bolts or by sprayed concrete.

Inundation, and collapse of the tunnel face. In tunnel construction with conventional timbering whenever even a trickle of water appears at the face it should be sealed. With full-face tunnel-boring machines the construction of the machine provides face support for most of the area. There is, however, a space in the front through which the conveyor gathers the spoil. Facilities need to be provided such that at the press of a panic button the conveyor is instantly retracted and a door closes the aperture.

Where inundation is a possibility it is vital that a properly planned escape route is maintained to allow all in the tunnel to be evacuated. Well-lit uncluttered walkways should be provided at the highest level practicable. Emergency lighting should also be provided.

In suspect ground, probing ahead should be undertaken to provide information about the need for ground treatment to prevent inundation. The equipment should include suitable non-return valves to prevent the possibility of a water inrush.

Falls from platforms and the danger from falling material. Hand excavation, timbering of the face, lining erection and other subsidiary work may be carried out from working platforms. The platforms should be properly designed and provided with close-boarded deckings, guardrails and toe boards to prevent operatives and materials falling from the working position. Scaffolding material is unlikely to be suitable for the construction of platforms near the working face as it is not sufficiently robust to withstand the continuous knocks from falling material.

Atmospheric pollution. The types of dangerous atmosphere that may be encountered are those due to

- oxygen deficiency (*see* pp. 63, 135, 136 and 162)
- carbon monoxide fumes, the usual source of which (other than fires) is the exhaust fumes of plant driven by petrol or diesel engines; petrol engines should never be used in tunnels during construction; compressors and any other static plant at ground level should be sited so that their exhaust fumes do not enter any access shaft, ventilation shaft or adit connected to the tunnel workings – it may be necessary to fit an extension pipe to the exhaust to convey the fumes away and down wind
- carbon dioxide, which may occur naturally, particularly in limestone areas, and is associated with oxygen deficiency
- methane, which occurs naturally and must be expected in coal measures, peat areas, carbonaceous strata and in or near refuse dumps (*see also* p. 63)
- hydrogen sulphide, which occurs naturally as a result of the decay of organic matter containing sulphur
- nitrous fumes, produced as a result of blasting operations and from welding.

Another atmospheric pollutant common in tunnelling is the respirable dust created by the excavation process and subsequent operations, dust which may be dangerous and may cause permanent damage to the lungs; moreover, the likelihood of bronchitis may be increased for those exposed to excessive concentrations of dust. The occupational exposure limits for all substances are given in guidance note EH40 published in Britain by the Health and Safety Executive. The types of dusts most likely to be encountered are

- silica (e.g. from sandstone, granite, flint and fossils), which is dangerous and may cause silicosis; another source may be spray concreting operations
- coal dust (from coal measures), which is dangerous and may lead to pneumoconiosis
- general dust (e.g. chalk, dry clay or cement), excessive exposure to which may lead to bronchitis; furthermore, fossils and flints (which are siliceous) may be found in chalk, for example.

The danger from atmospheric pollution may be averted or considerably reduced by the provision of an adequate system of ventilation (Figs 4.34 and 4.35).

Where dust is created by the excavation process, particularly where boom cutters are used, suppression by water spray may solve or substantially reduce the problem. Otherwise exhaust ventilation close to the source is required. Where the concentrations cannot be reduced below the recommended exposure limits appropriate protective breathing equipment must be worn. This may apply more often to those operations not directly concerned with the excavation process, e.g. spray concreting. Frequent routine atmospheric sampling and testing for the presence of dangerous levels of gases and dust likely to be encountered must be made.

When there has been a break in operations and the ventilation system has been shut down for several hours, for example over a weekend, a holiday or even after a breakdown in the ventilation system, the atmosphere over the full extent of the workings must be tested by a competent person wearing breathing apparatus. The atmosphere must be declared safe before any other person is allowed to enter.

Moreover, where particular operations, not necessarily continuous, such as spray concreting (shotcreting) and blasting are undertaken, specific atmospheric checks should be made. In the case of blasting, the checks should be made a few minutes after shot firing, and operatives should not be allowed back to the face to clear rubble etc. until the atmosphere has been declared safe. In all cases the necessary action to be taken will depend on the results of the tests and may require the evacuation of the workings until conditions are declared safe.

Further precautions include the following.

- There is a need to maintain effective communication between ground level and all working points in the tunnel. In relatively shallow shafts and short headings direct voice contact may be satisfactory or a simple sound signalling system by bell or whistle. In the larger and longer tunnels obviously more sophisticated systems are necessary. Frequent checks should be made from the surface to ensure that people in the workings are well and that the equipment is functioning correctly. This is particularly appropriate with small tunnels and headings where the majority of the work is at the face and carried out by a few operatives.
- Appropriate rescue and first aid equipment including suitable breathing apparatus should be kept at ground level ready for immediate use near the access shaft.
- In the event of someone in the workings being overcome by a dangerous atmosphere, entry for rescue must only be made by people wearing suitable breathing apparatus. In no circumstances should any rescue attempt be made by people not protected in this way.

Fig. 4.34. Good ventilation and lighting during tunnelling for the Jubilee Line extension for London Underground (courtesy Balfour Beatty Amec Joint Venture)

Fig. 4.35. Results of good ventilation during concrete spraying for the construction of A20 highway tunnels at Folkestone (note also the proper use of PPE) (courtesy Balfour Beatty main contractor)

Trapping. Where belt conveyors are used for muck disposal all dangerous parts such as chains, sprockets, gears and in-running nips – particularly between the belt and the head and tail pulleys – must be enclosed, or otherwise securely guarded. A continuous emergency stop switch cord should be provided along the whole length of the conveyor and be easily accessible. No maintenance work should be carried out on a conveyor in motion and an isolator switch should be provided for use when maintenance work is undertaken. There should be only one position from which the conveyor can be started. A distinct audible signal, understood by all who enter the workings, should be given, followed by a time interval to allow people to get clear, before every start-up. Riding on the conveyor belt should be prohibited.

The final positioning of all tunnel lining segments usually involves manual handling, and a great deal of care is necessary to avoid hand (or fingers) becoming trapped between units or between a unit and fixed equipment. Suitable hand tools for insertion in guidance holes will minimise this risk. In large tunnels where the segments are very heavy, special segment-erecting machines are employed but even with these the final positioning often requires manual assistance.

Transport. Where muck, material and workers are conveyed by rail transport on narrow-gauge track the main dangers are that people may be trapped and crushed between the waggons (or loco) and the sides of the tunnel or between waggon and waggon in the case of a twin-track system, collision with other plant or equipment in the tunnel and derailment.

Unless the size of the tunnel allows adequate clearance between the sides of the tunnel and the widest waggon (or loco), allowing for the swaying motion of the waggons, refuges should be provided at 18 m intervals along the tunnel to shelter people in safety while the train passes. They may be cut into the side of the tunnel or be prefabricated platforms fixed to the tunnel sides with guard rails and easy access. British Standard 6164[13] quotes 500 mm as the desirable minimum clearance between any part of a vehicle and any fixed equipment (that may be attached to the side of the tunnel). For the loco driver it is also important that there is adequate overhead clearance, which should be not less than 1·1 m from the seat of the loco.

Clear audible and visual warning of approach from the loco will enable people in the tunnel to take refuge and of course an efficient train braking system must be provided and maintained. The track should be of adequate section, properly supported, laid to line, level and gauge. It should have an even running surface and the rails should be properly jointed with bolted fishplates or other means. Poor track construction and maintenance together with excessive speed, especially on curves, are possibly the principal causes of derailment, and often the cause of collision with a fixture at the side of the track.

In small headings where muck skips may be conveyed on narrow-gauge track the maintenance of good clearances between the skips and the side is important for the safety of the timbering. A particular concern is the possible collision with a gas cylinder that may be for use in the tunnel. Storage of gas cylinders underground should be avoided and they should only be taken down for immediate use; even then they should be properly stored in a protected container well clear of any rail track.

It is vitally important for the loco driver to be trained and experienced and to be able to see ahead of the train. Thus both good lighting in the tunnel and a headlight on the train should be provided. Where the driver cannot see ahead of the train because of larger waggons being pushed in front of the loco, a banksman, in communication with the driver, should proceed on foot in front of the train to ensure clearance for the train to move forward.

In the narrow confines of tunnels great care must be exercised by non-rail transport drivers (e.g. dumpers) to avoid colliding with other vehicles and pedestrians. Again, good lighting and experienced drivers are essential for the safety of operations. The vehicles themselves should be fitted with good head and rear lights which are visible at 60 m, together with audible and preferably also visual warning devices indicating that the vehicle is about to move and a distinctive audible warning for reversing operations.

Noise. Noise should be kept to a minimum by the machinery design and by good maintenance. Where the noise levels cannot be reduced to an acceptable level the use of ear protectors should be used. Pneumatic tools, often employed in hand drives, may be substantially quietened by the use of muffle bags.

Electrical hazards. Electrocution and electrically initiated fires may be prevented by the proper design, installation and maintenance of all electrical equipment. Comprehensive guidance is provided by BS 6164 : 1990 *Code of practice for safety in tunnelling in the construction industry.*[13] In small headings where perhaps the only electrical equipment is the lighting, this is usually of the temporary festoon type. It should be supplied at a voltage not exceeding 50 V ac, with a maximum voltage to earth not exceeding 25 V ac Where for any installation there is a possibility of methane gas entering the workings only explosion-protected electrical equipment must be used.

Compressed-air hazards. The main hazards associated with tunnels driven in compressed air are

- decompression sickness, including aseptic bone necrosis
- increased risk of fire because of the greater concentration of oxygen
- inundation.

The likelihood of decompression sickness can be minimised by strict adherence to correct compression and decompression procedures. The Work in Compressed Air Regulations 1958 give rules for compression and decompression. However, more recently improved tables (rules) have been devised which may minimise the likelihood of workers suffering decompression sickness. They are known as the Blackpool Tables and may be used instead of the tables provided in the schedule to the 1958 Regulations.

However, permission to do this in Britain must first be obtained from the Health and Safety Executive.

The increased risk of fire calls for the highest standard of tidiness and housekeeping. The adoption of a fire-resistant fluid rather than mineral oil for hydraulic systems should be given careful consideration as the latter is highly flammable.

Burning and welding should only be carried out in the workings where absolutely essential. Even then it should only be allowed when authorised by the person in charge of the site. The work should then be strictly controlled and the following precautions should be taken.

- The bottles of flammable gas should be kept to the very minimum necessary for the task(s) and should be kept in workings for the minimum length of time.
- During the welding or burning operation all flammable material should be removed from the vicinity of the work.
- Any fixed equipment or material below the area of the welding or burning should be protected by a metal tray containing wet sand.
- Throughout the operation a member of the fire brigade should be in attendance equipped to extinguish any fire and should remain on duty, observing for the possible outbreak of fire, for at least 30 minutes after the completion.
- The gas bottles should be carried and stored in suitable protective containers to prevent damage from accidental impact and should never be placed where rail or road transport might cause damage from impact.

As a rule fire brigades will not provide fire-fighting and rescue cover for compressed-air tunnels. It is therefore necessary for the contractor to provide a specially trained fire and rescue squad. The squad must be fit to enter, and be experienced in, compressed-air work. Moreover, it is important that the squad remain familiar with the geography of the tunnel.

Except when the man lock is in use the inner door (to the workings) must be left open. Where the pressure exceeds one bar consideration should be given to the provision of decant locks because the man lock may become too hot to be usable throughout the time decompression period.

Failure of the air pressure will cause flooding. This may be due to plant failure or a blow at the face when the air escapes rapidly. For the first possibility it is necessary to have stand-by compressed-air plant; the compressor plant should be under the constant supervision of a competent operator able to switch air supplies in the case of main plant failure. It is a wise precaution if possible to provide air-tight steel, segmental diaphragms at intervals in the crown of the tunnel to form air pockets, where people can survive until the air pressure is restored.

Fire. A good standard of housekeeping is required in free-air tunnels as well as for those driven in compressed air. The amount of combustible materials and flammable liquids should be restricted to that necessary for each working shift. Flammable liquids should be properly stored in sealed screw-capped metal containers away from other combustible materials and away from positions where they may be inadvertently knocked.

Burning and welding should be carried out where possible above ground but when it is necessary in the tunnel only the minimum number of cylinders of the smallest size consistent with requirements should be taken into the workings. In free-air tunnels the cylinders should be carried in and stored as for compressed-air tunnels. For burning and welding operations in free-air tunnels the precautions required are the same as for compressed-air tunnels.

The assistance of the local fire brigade should be sought regarding the selection of fire-fighting and protection equipment, including a suitable audible alarm, and arrangements made for the brigade to help in case of fire. A fire and rescue plan should be evolved with the help of the fire brigade.

Safety audits
Regular inspection of the tunnel workings should be carried out by the safety supervisor to ensure that working procedures based on the hazard analysis are being adhered to and updated as necessary. The site manager should take urgent action to remedy any adverse findings.

Sewers and confined spaces

Hazards of sewer work
The hazards for those undertaking the inspection and maintenance of sewerage systems that are still in use, or have been in use, include

- injury from falls
- a sudden increase in the depth and flow of effluent
- dangerous atmospheres
- infection from bacteria.

Falls. Falls may occur at manholes when people are entering or leaving the sewer, or if they slip on the benching at the bottom of the manhole or on the invert of man-entry sewers while walking to or from the workplace.

In addition to the risk of abrasions and fractures from falling, drowning is also a possibility, even in a shallow depth of effluent, if the fall causes unconsciousness. Moreover, there is a particular risk of bacterial infection when the skin is broken by the fall. Injury may also be caused by tools or materials being dropped down a manhole and striking someone at a lower level.

Sudden increase in flow. Conditions in a sewer can change very quickly. A sudden increase in the rate of flow of effluent may sweep workers off their feet and lead to injury or even to drowning. This may occur following heavy rain, which could be either local or some distance from the area where work is being undertaken. It may also be due to high discharges from industrial premises or to the sudden release of effluent which has built up behind a blockage.

Dangerous atmospheres. Dangerous atmospheres may be due to a deficiency of oxygen, or to the presence of toxic gases or flammable or explosive gases. Oxygen deficiency can occur through its absorption by the sewage, or through the replacement of air by carbon dioxide produced by the decomposition of organic matter in the sewer.

Hydrogen sulphide is the most common toxic gas that is likely to be encountered in a sewer, and this is also a product of the decomposition of organic matter. It is not only very toxic but also very explosive. The highly explosive gas, methane, is also likely to be present in a sewer.

Leaks from gas mains and spillages or leaks from petrol storage tanks may find their way into sewerage systems and create flammable and explosive atmospheres. Moreover, discharges from industrial premises often contain substances which release toxic and explosive vapours.

Infection from bacteria. Infection from bacteria may result from skin contact with contaminated effluent. The sources include discharges from hospitals, research laboratories and industrial premises.

A particular occupational risk for sewer workers is skin contact with water that has been contaminated with the urine of rats. This causes Weil's disease which, unless diagnosed and treated early, can be very serious. It is advisable therefore that all sewer workers should advise their doctor of their occupation. Others who enter sewers for any purpose should also advise their doctors that they have done so. Infectious matter may also be introduced into the body through the mouth, nose or eyes, e.g. following a fall into the effluent (*see* p. 169 and Fig. 4.46).

Reducing the hazards of sewer work

Falls. The main cause of sewer workers falling within manholes is slippery, defective or loose ladder rungs or step irons. Ladders, step irons and landings should be hosed down periodically to clean off all silt, slime or other slippery substances. They should also be checked to ensure that they are secure and not defective; where they are defective the damage should be corrected.

The first person in a working party to descend a manhole should wear a harness with safety line attached which should be kept reasonably taut and controlled by a person at the surface. Until the first person signals from the bottom that entry is safe, no one else must enter except for rescue purposes. Thereafter, the rest of the party should enter with only one person at a time on the ladder. Moreover, in a man-entry sewer, they should move away into the sewer so as not to be directly below the person descending. Similarly, on leaving, only one person at a time should be on the ladder.

Manholes over 6 m in depth should be provided with rest platforms every 6 m and there should always be a rest platform in man-entry sewers just above the soffit level of the sewer. Preferably the platforms should extend over the full cross-section of the manhole and be provided with hinge trap doors. The platforms not only provide a resting place but also limit the distance of a fall and provide protection to those below from falling tools and materials.

All tools, equipment and material must be lowered (or raised) in a suitable container on a rope. They should never be carried by operatives on the entry ladder. When lowering or raising is being undertaken advance warning must be given from the top to allow people in the manhole to stand clear. Moreover, everyone working in sewers or manholes should wear a safety helmet.

The first person to reach the bottom of the manhole should secure a safety chain across the sewer on the downstream side. Furthermore, when the working area is reached, a safety chain should be secured across the sewer, immediately on the downstream side, to prevent workers being swept away should they slip.

When entry is to be made into a sewer, a safety rope, to be used as a handhold, or for the attachment of a lanyard from a safety harness, should be fixed in the sewer between adjacent manholes where steadying support cannot be obtained from the sides or when the effluent is particularly deep.

Sudden increase in flow. Before work starts the person in charge should ascertain whether any industrial establishments have reported their intention to make large discharges into the system. If so, entry should be delayed until the effect on the working area has passed.

Information about weather conditions for the area should be obtained and periodically updated and passed to the people at the surface. If a storm occurs in the vicinity or can be seen some distance away, the top person should signal to those underground to return to the surface immediately. Moreover, where those in the sewer notice an increase in the flow rate, hear the noise of approaching water or a distinct increase in air movement, they should leave the sewer immediately.

Dangerous atmospheres. It is essential to provide good ventilation. This can be done by removing sufficient manhole covers upstream and downstream (at least one either side) of the entry and exit manhole. Before the atmosphere is checked 15 minutes should be allowed to pass. No naked lights, smoking or any means of igniting should be permitted within 5 m of any removed manhole cover or, of course, in a manhole or the sewer.

The atmosphere in both the entry and exit manholes should be tested with appropriate gas detection equipment, and no entry should be made unless safe conditions are indicated. After the working party has entered the sewer the gas detection equipment should be taken with the party and frequent checks made of the atmosphere. Where conditions deteriorate the party must return immediately to the surface. Even when the gas detection equipment indicates safe conditions, if unusual smells are noticed, the party should return to the surface at once.

When entry is made into the sewer, the working party should keep close together and walk slowly and carefully to avoid disturbing the sludge since this may release toxic, flammable or explosive gases. Nylon material should not be worn and any boot studs must be of a non-sparking type.

Infection from bacteria. All cuts, scratches, abrasions and open sores should be protected by waterproof dressings before entry into a manhole or sewer. Moreover, any cuts or abrasions sustained in the manhole or sewer must be cleaned and treated immediately outside the workings. An application of a lanolin-based barrier cream before work starts and following the after-work wash reduces irritation and the risk of infection.

It is important to avoid contact of the skin with sewage and the face must not be touched with the hands. A person whose face or eyes become irritated should leave the workings, thoroughly wash and then receive attention. In sewer work, coveralls or sewer suits, sweatbands, gloves, studded thigh or wellington boots must be worn in addition to safety helmets and safety harnesses. Arrangements should be made for drying and cleaning the protective clothing.

At the completion of work it is important that all personnel should thoroughly wash their hands and forearms and clean their finger nails in hot soapy water containing an antiseptic. This must be done after protective clothing has been removed and before any food or drink is taken.

Safety at ground level. Safety barriers, cones and road signs must be erected around all open manholes, whether attended or unattended (see p. 124).

Communications. The underground working party must establish and maintain contact with the workers at the surface at all times, e.g. by telephone, radio or voice. They should also maintain visual and verbal contact with each other. A check call to ensure that the party below ground is safe and well should be made at intervals not exceeding 3 minutes from the time when the party leaves the surface. Calls may be verbal, or pre-arranged signals may be given on whistles or foghorns, e.g. one blast 'are you OK?'; reply, one blast 'yes'. When work is to be carried out along a sewer, someone should be stationed below ground at both the entry and exit manholes and the check call should then include the two workers below ground, i.e. surface worker (entry) to worker below ground (entry) to working party to worker below ground (exit) to surface worker (exit).

Size of working parties. In no circumstances should anyone enter a sewer or manhole over 5 m deep unaccompanied. Where it is necessary for a party to enter a sewer the minimum size of gang is seven; two surface workers and one worker below ground at the entry manhole, one surface worker and one worker below ground at the exit manhole and two in the working party. However, where work in the sewer is within 15 m of the entry manhole support workers at the exit manhole are unnecessary. Where the work is entirely within the manhole and its depth does not exceed 5 m the work below ground may be carried out by one person but with two surface workers in attendance.

Rescue from sewer. The local fire brigade should be informed of the work being undertaken and its location so that they are equipped and ready in case of an emergency. The workers at the surface should ensure that they know the location of the nearest working telephone or preferably a portable telephone or car phone should be available.

Rescue apparatus and equipment. The following rescue apparatus should be kept in the charge of the senior person at the surface in close proximity to the entry manhole

- first-aid kit
- two sets of rescue breathing apparatus
- one resuscitator
- one winch (ready for use)
- one escape self-rescue breathing set for each person below ground.

Fig. 4.36 shows training in the use of rescue equipment for an unconcious worker from a sewer.

Work in confined spaces

Every year there are a number of fatal and serious accidents because people carry out work in confined spaces on construction sites where the circulation of fresh air is restricted. Confined spaces include trenches, shafts, bored piles, boreholes, tunnels, headings, adits, sewers and drains, underground chambers, pits and ducts, pipelines, tanks, retorts, vessels, vats and silos, furnaces and ovens, box girders and columns, and closed and poorly ventilated rooms, particularly basements.

In 1996 the Health and Safety Executive issued a consultative document seeking the views of organisations on proposals for new Confined Spaces Regulations. With some exceptions they will apply throughout industry and will replace a number of outdated and conflicting existing regulations.

Hazards of confined spaces

The chief risks in confined spaces are those associated with flammable gases and toxic fumes and vapours and with the likelihood of there being insufficient oxygen to support life. These aspects are dealt with more fully in the sections concerned with excavations, tunnels and sewers (*see* pp. 62, 128 and 135). Reference may also be made to HSE guidance note GS 5 *Entry into confined spaces* (*see* Appendix 4).

Dangerous concentrations of gases and vapours can arise

- in an excavation, carbon monoxide or carbon dioxide or both from the exhaust of internal combustion engines
- from fissures in the ground, methane gas, especially in carboniferous areas, refuse dumps and reclaimed land with domestic and industrial waste; it can be flammable and highly explosive when mixed with air
- from the use of explosives, nitrous fumes

- in a vessel or container from dangerous chemicals remaining from a previous use
- when sludge containing dangerous chemicals is disturbed by cleaning operations
- in live sewers from industrial waste containing dangerous chemicals or biological action in rotting vegetation, for example
- when welding or flame cutting
- when using solvents or applying adhesives, e.g. waterproofing operations
- due to oxygen enrichment of the atmosphere by operations using oxygen, e.g. oxy-propane cutting.

Oxygen deficiency can occur in pits and headings, for example, as a result of the displacement of oxygen by carbon dioxide which has seeped in from the surrounding ground, particularly in chalk and limestone. Oxygen deficiency can also occur in steel box girders and in other confined spaces in steel structures where oxidation of the steelwork has occurred.

Safe systems of work in confined spaces

It is recommended that a permit-to-work system is followed for all operations that require entry into confined spaces (*see* HSE guidance note GS 5

Fig. 4.36. Training for rescue of an unconscious worker from a sewer (courtesy Shorco)

Entry into confined spaces and chapter 6). No staff or operatives should be allowed to enter a confined space if they are physically or mentally unsuitable. Rescue from confined spaces is difficult enough without the added burden of chronic disability.

Construction work in confined spaces is covered in generality in the MHSW Regulations (*see* p. 194) and in particular in the Confined Spaces Regulations (in preparation). In addition, the CHSW Regulations require that every workplace shall be ventilated by a sufficient quantity of fresh or purified air. If there is a likelihood that the atmosphere is not safe, no one must enter the confined space until the atmosphere has been tested by a trained competent person and until that person is satisfied that the air is safe to breathe. Atmospheric testing may be carried out by portable oxygen meters, gas detector tubes and explosimeters. Guidance is given in HSE guidance note EH42.

Confined spaces should be inspected by a competent person on every occasion before any person (other than the person carrying out the inspection) enters the confined space. Moreover within 24 hours of completing the inspection a report should be prepared giving among other things details of what could give rise to a risk to health or safety and any remedial action taken.

A safe atmosphere may be provided by the introduction of fresh air naturally or by forced ventilation and extraction. This may be by compressed air, either from cylinders or a compressor, by a blower fan and trunking and/or by an exhaust fan and trunking.

Where rigid trunking is used the fans can be reversed to dispel noxious fumes, for example where explosives are used in tunnelling. In all cases it is important that fans and trunking are maintained in good condition. Where the provision of fresh air is impracticable, people entering should wear breathing apparatus. The apparatus may be portable with the compressed-air cylinder carried on the back harness, or otherwise the cylinder should be located outside the confined space and the breathing apparatus connected to it by airline (*see* Fig. 10.7).

Working precautions

The precautions to be taken when working in excavations, tunnels and sewers are covered on pp. 63, 128 and 135. It is essential that trained personnel should be in attendance whenever someone is at work in a confined space and it is important to make arragements for dealing with emergencies and to inform all concerned accordingly. Emergency procedures should be tested from time to time. Those inside the confined space must keep in verbal communication with those outside or else rope signalling may be used. A portable telephone or two-way radio is desirable as an alternative. Rescue equipment must be provided and should include

- two safety harnesses, with an adequate length of safety line

- safe hand torches or cap lamps; an audible alarm
- a rescue team with breathing apparatus, first aid resuscitation stretcher and fire-fighting equipment.

Demolition and contaminated sites

Demolition

Demolition is one of the high-risk activities of the construction industry with a fatal and major injury accident incidence rate about 17 times that for the whole of the construction industry. Approximately 10% of all fatal accidents each year in the construction field occur in the demolition sector. Detailed guidance on acceptable standards and procedures in demolition is contained in BS 6187: 1982 *Code of practice for demolition*[14] and in HSE guidance note GS 29 *Health and safety in demolition work* (*see* Appendix 4).

Demolition is a very skilled process and, to be carried out safely and correctly, requires a great deal of knowledge of building construction and expertise in demolition methods. It is dangerously wrong to think otherwise (Fig. 4.37).

Key to safe demolition

For any demolition operation (including partial demolition such as in refurbishment projects) to be undertaken safely and efficiently, not only is a careful preliminary investigation of the structure necessary but also the

Fig. 4.37. Uncontrolled collapse of a multi-span brick arch bridge due to incorrect demolition (courtesy HSE)

results of each step in the demolition process must be accurately forecast. The steps to be followed are listed below.

Pre-survey inspection. A pre-survey inspection of the structure should be undertaken to ascertain

- the type of construction: this should include the contribution made by key elements to stability, e.g. by roof members, floors, walls and structural frameworks, etc.; detailed structural features should also be identified such as the counterweighting or holding-down method for cantilevered elements, tie bars – particularly where used in arch floor construction – the direction of main bars in reinforced concrete floors and the type of stressing in prestressed concrete members
- the state of the structure: the condition of structural members and of the fabric; damaged areas, previous repairs and alterations should also be noted
- the need for temporary shoring or other support: this is particularly relevant for attached buildings, arch construction and refurbishment
- special features: this would include the presence of dangerous materials or chemicals, e.g. asbestos, lead paint or dust, flammable residues and vapours in storage tanks or process vessels and contaminated ground
- the position of services: it is important to identify the whereabouts of gas, electrical and water services and to ensure that they are disconnected or isolated before demolition work starts.

CDM Regulations. Useful information to assist with the pre-survey inspection will be obtained from the health and safety file held by the client as a duty under the CDM Regulations for all structures to which these regulations apply (*see* p. 20).

Demolition method. The pre-survey should identify the problem areas and provide all the information necessary to determine the method of demolition to be used. The object should be to limit the number of people at risk and thus to adopt where possible methods which make it unnecessary for people to work at heights.

Moreover, as already indicated, it is very important to forecast the result of each step in the demolition process and thus to determine the precautions necessary to prevent premature collapse causing injury to personnel.

Method statement. When the method of work has been determined it should be set down in detail in a clear and understandable form. It should include

- the method to be adopted and the sequence of operations, the plant and equipment to be used and the details of working platforms and access routes required
- where any pre-weakening is to be carried out, the precise details of how this is to be done

- arrangements for the protection of the public and also for all personnel employed on the site; the arrangements should include the provision of protective fans or nets, and hoardings or other provision for preventing public access to the building being demolished; for site personnel the arrangements should cover the operation and maintenance of all plant and the provision and use of the necessary personal protective equipment
- arrangements for the disconnection or isolation of services
- in detail, the methods to be used for dealing with dangerous materials and hazards chemicals, e.g. asbestos and residues in storage tanks
- the identity and responsibilities of the person in charge of operations and the responsibilities and duties of people to whom control of certain aspects of the demolition work will be delegated.

Site supervision. Obviously not everyone engaged in demolition work will understand the structural significance of the various component parts he has to demolish and therefore demolition workers will need close supervision and guidance from someone skilled in the work if they are to proceed in safety.

It is important that supervisors have a considerable knowledge of building construction as well as demolition methods. They must always be experienced and competent in the particular operation being undertaken and be able to provide step-by-step guidance for each stage of the operation.

The CHSW Regulations require that anyone planning or carrying out demolition or dismantling work must be under the supervision of a competent person

All concerned with site supervision should be familiar with the method statement and should have access to a copy that should be kept on site.

Safe clearance zones. A restricted zone for all people should be maintained around the structure to be demolished. This should be a clear distance of 6 m or half the height of the structure (whichever is the greater) from the point of fall so that the debris can fall freely. If this is not possible then the debris must be lowered in skips or by chute.

When mechanical plant (e.g. cranes, grabs, pusher arms or lifting machines for balling purposes) is employed on site, a minimum restricted zone 6 m wide must be maintained from the face of the building. When the demolition method is pulling by wire rope, the restricted zone should be three-quarters of the distance between the winch and the structure to be demolished on either side of the rope and forward of the winch (or tracked vehicle) used for pulling. The appropriate restricted zone when tall structures such as chimneys are being felled should be an area of radius $1\frac{1}{2}$ times the height of the structure measured from the base.

Methods of demolition

The danger in demolition work varies with the method employed, which itself has a bearing on the number of workers at risk and their working time

Fig. 4.38. Unacceptable method of demolition (courtesy HSE)

at heights and in and around the structure being demolished. Methods of demolition may conveniently be divided into three categories, hand, mechanical and explosive.

Demolition by hand. Demolition by hand is progressive demolition, carried out piecemeal with hand tools such as sledge hammers, jack hammers, concrete breakers and cranes for lifting and lowering purposes. It is carried out in reverse order to the construction process, i.e. the roof covering is taken off first, followed by the roof trusses and upper ceiling and the building is then demolished storey by storey.

When this method is used properly protected working stages should be provided where necessary. Standing on the wall being demolished is unnecessary and should not be permitted (Fig. 4.38).

When operatives are expected to work at unprotected edges, such as when breaking up a floor slab, harnesses should be worn with appropriate safety lines suitably anchored. Where parts of floors have been removed to allow debris to fall through, access to the areas concerned should be prevented or guard rails and toe boards erected at a distance from the edges to protect workers from falling debris. Similarly, any holes in floors in the building under demolition should be protected by guard rails and toe boards to prevent people falling.

Structural collapse has occurred where floors have been overloaded by the accumulation of debris from parts of the building already demolished. Debris should not be allowed to build up, but should be discharged

Fig. 4.39.   Use of purpose-made man-riding skip to provide a safe working place for demolition of a high structure (courtesy HSE)

regularly by chute or skip or by dropping it to ground level. Where debris is dropped internally the build-up at ground level must be limited to avoid imposing excessive loads on the walls. No debris should be allowed to fall while clearing operations are in progress.

Where scaffolding is used to provide a working platform care is needed to ensure that sufficient ties are provided at the lower levels before demolition (and hence removal of ties) at the higher levels. Work on roofs should only be carried out from properly constructed platforms or crawling boards. Safe access for demolition workers is sometimes difficult to provide. Man-riding skips (Fig. 4.39) may be used.

Mechanical methods of demolition. Mechanical methods of demolition include the use of a demolition ball or pusher arm; inducing deliberate collapse; and pulling by wire rope: some examples are given below. The great advantage of mechanical methods is that they considerably reduce the number of people at risk, i.e. the number of people required at height on the structure. Sometimes it is necessary to demolish by hand until the structure has been reduced to a height convenient for one of the mechanical techniques to be used.

For demolition by *demolition ball* a swinging weight is used suspended from a lifting appliance, to demolish the structure progressively. The method should only be carried out by skilled operatives. Before a machine is used for balling operations the advice of the machine manufacturer should always be sought to ensure that the machine is suitable for the work.

Great care is needed to avoid overstressing the lifting machine and it is preferable for the ball to be dropped vertically or swung in line with the jib. Where a slewing action of the jib is used the angle of slew should not exceed 30° and the acceleration of slew and rate of checking should not be excessive – advice should be sought from the machine manufacturers.

The angle of the jib should not exceed 60° to the horizontal and its head should never be less than 3 m above the part of the building being demolished. When this technique is used the floor areas at all levels should first be reduced by 50–75% to allow for falling debris.

For demolition by *pusher arm* a steel attachment is used, fixed to an extended jib arm on mechanically operated mobile plant. It exerts a horizontal thrust similar in technique to a battering ram and is used to demolish walls.

The equipment should be on firm and level ground and used from outside the building. The cab needs to be protected against the impact from falling debris and the windows should be made of shatterproof glass. The application of the pusher arm to the wall being demolished should never be more than 600 mm below the top of the wall.

In the demolition method of *deliberately induced collapse* key structural members are removed, causing the whole or part of the building to collapse. The method requires a high degree of planning and the use of expert engineering advice. Where partial demolition of a structure is to be carried out by this method consideration must be given to the effect it may have on the section to be left standing. Where there is the slightest doubt about the stability of the remaining part the method should not be used.

Collapse is usually brought about by pre-weakening the key members and their final removal by means either of explosives or the method of pulling by wire rope and winch (or suitable vehicle). The most dangerous part of the method is the pre-weakening when, by the very nature of the exercise, operatives are working on and below a structure close to instability. The planning and weakening operation must be such that wind loads, likely vibration from work being carried out in the vicinity and any impact loads (e.g. those which might be caused by the operatives carrying out the work either deliberately or accidentally) must not tip the balance to instability.

Structural steel elements are usually weakened by cutting with an oxyacetylene torch and the type (pattern) of cut is important. Fatal accidents have occurred through incorrect cutting and overweakening of the supporting pattern. With masonry or reinforced concrete buildings a series of rectangular holes is cut towards the base so that the remaining building is effectively held up on a number of masonry (or concrete) columns – between the holes. The holes are cut with hand tools – jackhammers etc. – and care must be exercised to ensure that falling masonry from the top of the holes does not cause injury; temporary support may be necessary, especially on fairly thick walls where the operatives may have to stand within the cut area.

Before collapse is implemented all personnel should retire to a safe distance, and where explosives are used be protected from possible flying debris by a suitable barrier.

In addition to its use in the final stages of the deliberate collapse method of demolition, *pulling by wire rope* may also be used for parts of buildings such as chimney stacks and walls without pre-weakening, other than perhaps partial removal by hand of the adjacent walling and ceiling or floor.

The wire rope is fixed around the wall (or part of the building to be demolished) at one end and the other end is fixed to the pulling winch or tracked pulling vehicle. The main dangers are that people, in particular the winch operator or driver, may be struck by falling material; the rope may sever and snake, hitting either the operator or people in the vicinity; where the method is unsuccessful and the part being pulled remains standing, it and adjacent parts will have been weakened and may be in a dangerously unstable state; moreover, there is the danger of the pulling vehicle overturning if it lacks weight and power for the job.

The method of pulling by wire rope should not be used on buildings (or parts) that are more than 21 m in height. The length of the rope should be such that the winch operator or driver is positioned at a horizontal distance from the building of not less than twice the height of the highest part being pulled. Measures must be taken to ensure that people are prevented from entering the restricted zones indicated in the method statement, and the winch operator or driver must be protected from falling or flying debris by a suitable substantial cabin.

Steel wire ropes should be used, in good condition (undamaged) and of minimum circumference not less than 38 mm. However, the correct size of rope will depend on the load imposed by the operation. The rope should be examined by a person competent and experienced to judge the quality, condition and adequacy of ropes for the work, before use, at least twice daily and again before further use where damage during use is suspected. The rope should be protected where it is wound around sharp edges to prevent high stress points developing and to reduce wear.

The winch or pulling vehicle should be aligned in the direction of pull. Where a winch is used it should be well anchored and where a vehicle is used it should not lift off its tracks – if this starts to occur the operation must be abandoned.

When the operation starts the load should be gradually applied – snatch loading should never be used. Should the structure not collapse following the application of full pulling load an alternative mechanical method must be used. In no circumstances should hand demolition be reverted to nor should any operatives be allowed on to or even to approach the structure. The most suitable alternative methods are the use of a demolition ball or pusher arm.

Other mechanical methods include a number of specially designed hydraulically operated devices for demolition without the need for operatives

Fig. 4.40. Safe mechanical demolition using machine-mounted impact hammer (note the workman not wearing a safety helmet – this is illegal and dangerous) (courtesy HSE)

to work at heights. They include machine-mounted impact hammers (Fig. 4.40), power shears (an alternative to oxy-acetylene cutting)(Fig. 4.41), grabs, nibblers and concrete bursters (also gas-expansion concrete and masonry bursters). The main dangers are from falling material, which is overcome by the maintenance of restricted zones for personnel and cab protection for the operator where applicable; and overloading of the machine causing collapse of the arm or even overturning of the machine, which is overcome by the use of trained and experienced operators.

Use of explosives in demolition. Given the necessary expertise demolition by the use of explosives is possibly the safest method of demolition because the fewest people are at risk. Only specialists with the required experience and training should plan and control this type of work (Fig. 4.42).

The main danger is from flying debris; accidents may be averted by

- use of the correct type and charge weight of explosive: this is not usually a problem when an explosives expert is in control
- use of adequate blast mats to reduce the effect of air blast and flying debris
- the provision and maintenance of a restricted zone around the operational area, which should be decided by the experienced person in charge of the operation; before the charge is fired everyone except the

Fig. 4.41. Safe mechanical demolition using power shears for steelwork demolition (courtesy HSE)

shotfirer should be excluded from the area and not allowed back until the all-clear signal is given; the shotfirer should be protected from flying debris by a suitable barrier
- the establishment of a firing timetable together with an audible and visual signalling system to alert everyone, both site personnel and the public, of the intention to start firing; the visual arrangements should include the use both of red flags (to indicate danger) and notices at intervals along the border of the restricted zone and about the site indicating the work to be undertaken, the dates and times of firing, the audible warnings to be given before firing and the significance of red flags flying
- use of sentries: before the charge is fired sentries should be posted at

Fig. 4.42. Carried out by experts, the safest method of demolition is by explosives. The figure shows a vertical drop on a confined site. Despite rigorous checking that the danger zone was clear before blowdown, a member of the public with a small child on his shoulders is seen on the adjacent roof in this press photograph (courtesy W.J. Hatt Ltd and Reading Chronicle)

intervals around the perimeter of the restrictive zone and at all road and track access points to prevent entry.

Other precautions which need to be taken to prevent accidents when explosives are used in demolition operations are as follows.

- Correct storage and transportation of explosives and detonators is important: in particular, detonators should be kept and carried separately from explosives until they are required for use and should only be handled by those authorised to do so.
- Neither matches nor lighters should be carried by people handling explosives or detonators or by people in or close to storage places except when needed for firing a charge.
- Explosives should not be handled or kept near the path of vehicles or other mobile plant, where there is danger from falling material; nor should explosives be handled or kept close to sources of heat. Electric detonators should be kept well away from electric welding equipment, electrically operated machinery and vehicles with starter motors, generators and batteries, etc., and from radio and radar equipment.

- Except for the delivery of explosives all vehicles should be prohibited from entering the blast area.
- Electric detonators should only be carried around the site in boxes made of non-conducting material.
- The number of detonators used against those issued should be recorded and the detonator box kept locked until the detonators are needed.
- After the charge has been fired the all-clear should not be given until the shotfirer and person in charge of the explosives operation have conducted an examination of the area and satisfied themselves that it is safe, particularly after a misfire. Where a charge fails to explode no approach should be made for at least five minutes.
- All mobile plant should be moved to a place of shelter (from flying debris) or to a point outside the restricted zone before firing occurs.
- After the charge has been fired, should the structure remain standing all personnel must be prevented from entering the restricted zone until an assessment has been made; the competent person in charge of the operation, with the assistance of other experts (e.g. a structural engineer), as necessary, should determine the further measures required for complete demolition. Moreover, they must assess what (if any) safety precautions need to be taken, e.g. the erection of barriers with notices to prevent access.

Further guidance on explosives is given in BS 5607: 1988 *Code of practice for the safe use of explosives in the construction industry.*[15]

Health hazards from demolition

Health hazards from demolition processes arise from

- inhalation of toxic gases, vapours, fumes and dust
- ingestion of corrosive or toxic liquids and dust
- reaction with or absorption by the skin of corrosive, toxic or irritant dusts, powders, liquids and chemical substances
- exposure to high noise levels
- exposure to vibration.

Whenever exposure to hazardous substances is suspected, employers must carry out a COSHH assessment and re-assessments as necessary. Where health risk is likely the procedures laid down in the COSHH Regulations must be followed (*see* p. 158).

Where there is danger of inhalation of toxic gases etc. a method of work should be devised to minimise the generation of dust and fumes, e.g. in the case of dusts by damping with waterspray and by containment with chutes, covers, etc. Where the risk from dust or fumes cannot be completely eliminated, appropriate personal protective breathing equipment must be used by all those exposed to the hazard (Fig. 4.43).

Fig. 4.43. Use of proper protective clothing and breathing apparatus during asbestos removal (courtesy George Wimpey & Co. Ltd)

It is also important to ensure that airborne contaminants are prevented from escaping beyond the site and affecting the general public. Environmental monitoring should be carried out in the working area and at intervals around the site boundary where toxic dust and fumes, and excessive nuisance dust or both are likely to be generated.

With regard to the danger from ingesting and absorbing toxic substances, not only should personal protective equipment be used but also it is most important that a high standard of personal hygiene is practised. Thus adequate washing facilities must be provided at the site.

With regard to the danger from exposure to high noise levels, operatives should not be subjected to continuous noise levels over 90 dB(A). Where noise levels are higher than 90 dB(A) or where peak levels exceed 200 Pa ear protectors should be supplied and worn. Furthermore protectors must be supplied to employees who are exposed to 85 dB(A) per eight-hour day if they ask for them (*see* p. 164). The most likely sources of high noise levels are concrete breakers and compressors.

Initially, noise reduction by engineering means (e.g. surrounding concrete breakers with sound-absorbing bags) should be carried out to reduce the noise level to as low a level as practicable before the use of ear protectors is resorted to. Moreover, exposure time should be kept to a minimum by means of job rotation for operatives.

With regard to the danger from exposure to vibration, concrete breakers and pneumatic drills are examples of types of plant, widely used on demolition work, which may give rise to vibration white finger. The use of gloves and the device of restricting exposure by means of job rotation may reduce the risk of serious disability.

Examples and sources of toxic dust, fumes, etc. are as follows.

- *Asbestos*-containing material has been used for both thermal and acoustic insulation purposes. It may be found around hot vessels and pipelines, sprayed on the underside of roofs, and on walls and columns of buildings and structures. It may also be found in sheet form, i.e. asbestos cement, as roof covering and partition walls and in flooring material. All demolition operations will create toxic asbestos dust and it is important to realise that no level of exposure, for any type of asbestos, has been accepted as being safe.
- *Lead* dust may arise from the cold cutting of elements covered with lead-based paint or when metallic lead is handled. Lead fumes may arise from the flame cutting of metalwork painted with lead-based paints for example, and from the dismantling of tanks containing lead-based petrol.
- Flame-cutting plant may contain chemical deposits of *zinc* or *cadmium* and steelwork may have been painted with *zinc* or cadmium paint.
- *Silica* may occur in natural stone (e.g. sandstone), some bricks and concrete aggregate, and any method of demolition of structures constructed from these materials, will give rise to dust containing silica.
- *Ammonia, chlorine* and *aniline* are examples of gases and fumes that may be released during the demolition of refrigeration and chemical plant unless they have been properly purged before the dismantling demolition.
- Various *toxic sludges* may be encountered during demolition of virtually any storage tank forming part of an industrial process.
- All storage tanks and reservoirs in industrial premises may contain harmful products such as *corrosive liquids*.

A number of HSE guidance notes on health hazards that may be associated with demolition are listed in Appendix 4.

Working on contaminated sites

Increasingly, civil engineers, builders and demolition workers are required to design or construct works on sites that are contaminated by waste products. Contamination occurs from domestic and industrial refuse and from processes involving the production and storage of chemicals.

Many chemicals are injurious to health if absorbed into the body; some show immediate effects and others are serious only if built up over a period. Absorption may be through the skin, inhalation or ingestion. Examples of chemical hazards and precautions to be taken are given on pp. 159 *et seq.* Sites where contamination might be expected include

- asbestos producers
- chemical plants and depots
- dockland areas
- explosives factories and depots
- gas works
- landfill sites of domestic and industrial waste
- metal smelting and refining plants
- metal treatment and finishing works
- mines and quarries
- oil production and storage depots
- paints and graphites factories and depots
- railway yards
- scrap yards
- sewage works
- steelworks
- tanning and associated trades
- water treatment plants.

The distribution of contaminants is often random and not always identified by trial pits and boreholes. The contaminants may migrate through the ground with ground water and their chemical compositions may change by contact with one another. Some organic substances decompose, producing methane and sometimes hydrogen sulphide (*see* pp. 63, 128, 135, 136 and 162).

Identification and remedial action. The likelihood of contamination on a site may be confirmed by reference to clients' records, local authority records, old maps, etc. Soil samples should then be obtained, taking care not to expose to risk those carrying out the task, and sent to a suitable laboratory for analysis and estimation of the concentration of harmful chemicals present. No construction or demolition work that might endanger people on or near the site should be allowed until the contaminants have been identified and precautions taken.

For the development of contaminated sites it may be necessary to remove some or all of the contaminated material. Disposal of contaminated material comes under the Control of Pollution (Special Wastes) Regulation 1980, which is administered by the waste disposal department of the local authority. This will influence the cost of disposal and must be taken into account in the preparation of tenders. There must be no encouragement of fly-tipping. Where removal is considered unnecessary or undesirable the site may be covered by a blanket layer of granular material, the thickness depending on the protection required, the topography and future construction on the site, e.g. piles, foundations and drain or service trenches.

General advice on the problems of contaminated sites may be obtained from the Interdepartmental Committee of the Redevelopment of Contaminated Land, Department of the Environment. Advice may also be obtained from the local office of the Health and Safety Executive. A number of consultants specialise in the monitoring, measurement and control of contaminants and are able to advise on removal and replacement or on other site surface preparation.

Advice may also be obtained from the CIRIA Special Publication 108: 1995 *Remedial treatment for contaminated land*. The report is in twelve volumes which collectively provide a systematic approach to the management and remedial action for contaminated land. Safety is an important and consistent theme with each volume. Individual sections refer as appropriate to the guidance available from the Health and Safety Executive and to complementary CIRIA reports giving guidance on safe working practices on contaminated sites and on the use of substances hazardous to health.

Site precautions. People at risk include surveyors, site investigation teams, construction workers, demolition workers, visitors to the site and

Fig. 4.44. Simple wheel-washing facility to remove contaminants from vehicles before leaving the site

others in the vicinity. Suitable warning notices must be displayed and secure fences erected.

Wherever site contamination is suspected, employers have a duty to carry out a COSHH assessment and re-assessments as necessary, and to follow the procedures laid out in the COSHH Regulations (*see* p. 158). In addition to COSHH there are more specific regulations which sometimes will apply where lead and asbestos are found. Guidance on the Control of Lead at Work Regulations 1980 and the Control of Asbestos at Work Regulations 1987 is available from HSE Publications (*see* Appendix 4).

Workers on the site should be provided with

- a detailed work plan
- information on the hazards and precautions to be taken, and advice on personal hygiene
- clean overalls, gloves and other protective wear as necessary
- a site hut with toilets, washing and changing facilities and a suitable eating place.

When work is being undertaken on contaminated land it is important that vehicles leaving the site do not carry contaminants away on their tyres, mudguards and underbodies. Members of the public and workers elsewhere may be affected by contaminants discharged on highways or other site roads. This risk should be avoided by the installation and use of a wheel and body washing facility (*see* Fig. 4.44).

Work over water

For work over or in the vicinity of water, the ever-present risk of falls is compounded by the risk of drowning. In the prevention of accidents the approach recommended here is to reduce the possibility of falls and to provide rescue facilities and equipment.

Prevention of falls

Working platforms, places and access gangways to them should be properly constructed to the widths required by the CHSW Regulations. Rigid guard rails and toe boards, or barriers or other protection, should be fixed at the correct height at all edges from which falls are possible. This also applies to the edges of openings within the structure through which falls are possible, e.g. the uncompleted formwork of a bridge deck.

The decking to workplaces and gangways should be close boarded and effectively secured to prevent displacement by strong or gusty winds or by rising water from tide or wave action when the decking is close to the water surface. Ladders for access should be sound, securely fixed and of sufficient length to enable people to get on and off without difficulty.

Tripping and slipping hazards must not persist unchecked. Unnecessary materials, tools and equipment should not be stored or left at working places or on gangways. Materials required for the work should be restricted to those for the immediate tasks in hand and should be neatly and securely stacked. The decking of working places and gangways in marine work can become wet, slimy and greasy. The surfaces should be regularly cleaned, e.g. to prevent slipping. Sand that has been put down for cleaning oil spillages should be swept up. During periods of frost or snow, cleaning and salting should be carried out and the salt should be swept up before other work is carried out.

Non-slip footwear should be worn by everyone using platforms and gangways. Where the provision of safety nets immediately below the working areas is impracticable, the use of safety harnesses with lanyards continuously attached to a secure anchorage is advisable.

Whenever visibility is not good, a high standard of artificial lighting must be provided for work areas and accesses. This should also illuminate the surface of the water adjacent to site operations.

Rescue equipment

All people working over water should be provided with and encouraged to wear a brightly coloured life jacket, or other buoyancy aid capable of bringing a person to the surface, even if unconscious, in a face-up position. It should have self-igniting lights to facilitate easy identification in conditions of poor visibility. Buoyancy aids should comply with BS EN 394: 1994 *Life jackets and personal buoyancy aids additional items* and BS EN 396: 1994 *Life jackets and personal buoyancy aids. Life jacket 150*[16] or be of a type approved by the Ship and Boat Builders National Federation.

Board of Trade approved life-buoys with lifelines should be provided at intervals along the site and positioned conveniently for use in an emergency. In addition, lightweight emergency lifelines should also be readily available. They can be thrown up to 40 m with reasonable accuracy and are easily recoverable for rethrowing when necessary.

Grab lines should be attached at intervals to structures or floating plant. A marker float should be attached at the free end. It may be prudent to provide additional grab lines downstream of the operations. Checks should be made at the beginning of each shift to ensure that rescue equipment is present and in good order.

Rescue facilities

It is essential to provide arrangements for the prompt rescue of anyone falling into the water. Moreover, at every place where work is being carried out on the structure at least two people should be employed so that, should one fall, the other can raise the alarm.

A suitable rescue boat manned by two experienced boat handlers wearing life jackets and who are also qualified first-aiders must continuously patrol the site of operations while work is in progress. The boat should be fitted with grab lines, at least one life-buoy or lightweight emergency lifeline, a boat hook, anchor and line and a baler. Two-way communication with the site office and the working level should be provided. Where work is to be carried out in darkness or at dusk, a swivelling searchlight mounted on the boat is essential. For rescue purposes it is useful to keep on board a pole with a hooked end.

Water transport

Where people are conveyed by water to and from their place of work the boats provided must meet the requirements of the Merchant Shipping Act. Where the craft is intended to carry more than 12 people in addition to the crew, it must be surveyed at least once a year by a Board of Trade surveyor and certified as being seaworthy, safe and possessing the required equipment. This includes inflatable life rafts, life jackets, life-buoys, distress signals and fire extinguishers. There are no rules for craft smaller than for 12 passengers but, of course, as with larger boats, the crew must be competent and experienced boat handlers.

HEALTH HAZARDS

Legislation

A number of Acts and regulations that have been concerned with the health of workers were brought together and their scope extended when the Control of Substances Hazardous to Health Regulations came into force in 1989. These regulations specify what is to be done by employers and others to satisfy their general duties to maintain a healthy working environment and safe systems of work under the HSW Act 1974.

Control of Substances Hazardous to Health (COSHH) Regulations 1994

The aim of the COSHH Regulations is to maintain and improve existing standards of health and safety by protecting people against risk to their health, whether immediate or delayed, from exposure to harmful substances.

The regulations apply to substances that are classified as toxic, harmful, corrosive or irritant and to those which have maximum exposure limits and those with occupational exposure standards (OES). They also cover substances that have chronic or delayed effects. The meaning of substances hazardous to health is given in Regulation 2.

It is the duty of every employer not to expose employees to hazardous substances unless an assessment has been made of

- the risks to health created by the work
- the steps to be taken to achieve adequate control of exposure
- identification of other action necessary in respect of control measures, monitoring the workplace, health surveillance and providing information, training etc.

An assessment should involve

(*a*) consideration of:
  ° which substances employees are liable to be exposed to
  ° what effect those substances can have on the body
  ° where the substances are likely to be present and in what form
  ° the ways in which employees or other persons could be exposed
(*b*) an estimate of the exposure
(*c*) where valid standards exist representing adequate control, comparison of the estimate with these standards.

Every employer must also

- prevent or control exposure
- use, maintain, examine and test the control measures
- monitor exposure at the workplace
- carry out health surveillance
- provide information, instruction and training.

The duty imposed upon the owner or occupier of the premises is to carry out the above measures for the protection not only of his own employees but also of others at the premises or likely to be affected 'so far as is reasonably practicable', except that he is not responsible for health surveillance of employees not his own. This is specially relevant to civil engineers and others in the construction industry. For employees working on sites or establishments where there are hazardous substances the onus is upon the employer to provide the required health surveillance of his employees. Duties are also imposed upon the self-employed, except for monitoring exposure and health surveillance.

A useful publication is HS (G) 97 *A step by step guide to COSHH assessment* (*see* Appendix 4).

Full details of the regulations and advice on how to comply with them are given in the approved codes of practice *Control of substances hazardous to health and control of carcinogenic substances* available from HSE publications (*see* Appendix 4).

Chemical hazards

Dangers from chemicals may arise from contact with the skin, inhalation or ingestion (swallowing).

Contact with the skin

Dermatitis is perhaps the most common occupational skin disease, and can be caused by contact with a number of materials commonly used in the construction field. Cement, particularly when wetted, is an example and if it contains chromates the danger is increased. The slaked lime content in cement is considered to be responsible for cement burns on the legs and feet of operatives where concrete has spilled into their wellington boots.

Certain epoxy resins used in grouts, seals and adhesives are another cause of dermatitis, as are acids, alkalis, solvents, thinners, paints, varnishes, acrylic and formaldehyde resins, brick and stone dust, pitch, tar and bitumen.

To avert this hazard, it is clear that skin contact with chemicals should be avoided. Careful handling, including the prevention of splashing, is called for, together with the use of closed stout containers for distributing, and appropriate tools for mixing and applying the products. Protective clothing should be worn, in particular suitable gloves and safety eyewear. Barrier creams on the hands and forearms may be of use but care is needed to select the correct product for the particular operation.

Personal cleanliness is most important and hands must be clean before eating, smoking or using the lavatory. Gloves should be thoroughly washed before they are removed. An individual may become sensitised (i.e. allergic) to a particular substance by repeated exposure and even after medical treatment the dermatitis will recur on further exposure to the substance.

Inhalation of harmful chemicals

The inhalation of certain dusts, fumes, gases and vapours may give rise to a variety of risks to health, including respiratory disorders, poisoning, asphyxiation and cancer. HSE guidance note EH 40 (*see* Appendix 4) gives occupational exposure limits for many airborne contaminants. It is revised annually. Some of the more common substances are now discussed.

Asbestos dust. All types of asbestos dust are dangerous and breathing them may lead to asbestosis, which is a scarring of the lungs causing shortness of breath. Other hazards are lung cancer and mesothelioma, a rare cancer of the lung envelope. Asbestos is present in a number of products used in construction processes, e.g. asbestos cement roofing sheets, decking, cladding sheets, rainwater and soil pipes. Their cutting, handling and breaking gives rise to this dust. It is also found in asbestos insulating board and in sprayed coating for insulation and fire protection purposes. (This process using asbestos-containing material is prohibited under the Asbestos (Prohibitions) Regulations 1985.) However, stripping out the coatings continues to be an operation very hazardous to health. The Asbestos (Licensing) Regulations 1983 are intended to prevent employers and the self-employed carrying out work that involves disturbing asbestos insulation and coatings unless they are licensed to do so by the Health and Safety Executive.

Wherever possible safe substitute materials should be used for the building process. Otherwise, exhaust ventilation equipment, approved respirators and appropriate protective equipment must be used when handling and manipulating asbestos or asbestos-containing materials. For operations such as drilling, cutting and cleaning up, the area should be dampened and the operation carried out carefully to minimise the creation of dust.

A particular risk associated with dust on protective clothing is that the dust may be breathed in by anyone handling the garments, e.g. for cleaning. Immediately after use, protective clothing should be carefully placed (so as not to cause dust) in a sealed container appropriately marked 'asbestos- containing clothing' and sent for cleaning by cleaners competent to carry out the work. Those intending to work with asbestos should refer to the Control of Asbestos at Work Regulations 1987 and guidance notes published by the Health and Safety Executive (*see* chapter 8 and Appendix 4).

Silica dust. Silica dust is another commonly found dust which causes damage to the lungs. The lung scarring is called silicosis and causes breathing difficulties similar to asbestosis. Free silica is present in many rocks, aggregates and sands and danger arises from carrying out processes involving such materials. The processes may include tunnelling through rock, slate or ground containing flint nodules; the abrasive cleaning of stone structures, grinding, cutting and working granite and other stone; rock drilling and blasting. The precautions to be taken include standard dust-control methods such as exhaust ventilation, wetting down and, where appropriate, the use of protective breathing equipment and dust-proof protective clothing.

Carbon monoxide. Carbon monoxide is one of the most dangerous gases likely to be encountered on a construction site. It is odourless and, therefore, easily breathed by the unsuspecting. Headache, nausea, dizziness and loss of co-ordination may result from low levels of exposure, whereas high levels lead to rapid death. Moreover, regular low exposure may lead to heart disease.

Sources include internal combustion engine exhausts, poorly maintained space heaters (with incomplete combustion of the fuel), coke braziers and carbon dioxide welding. The risk is particularly high in confined working areas such as tunnels and excavations. In all cases the need for adequate ventilation is paramount to avoid danger.

Where carbon monoxide is likely to be present then appropriate monitoring of the atmosphere must be carried out.

Carbon dioxide. Carbon dioxide occurs naturally in chalk and limestone areas. High concentrations in confined spaces result in oxygen deficiency and very dangerous atmospheres. Numerous fatal accidents have occurred when unsuspecting workers have entered manholes in surface-water drainage systems or boreholes in such areas.

In all shafts, wells, adits, tunnels and excavations in limestone and chalk areas it is particularly important before each entry and continuously during all working time that the atmosphere should be tested for high concentrations of carbon dioxide and for oxygen deficiency. A further most important check is the barometer reading. If it falls during the time workers are in the excavation the atmosphere may change lethally, and therefore immediate oxygen-deficiency testing should be undertaken. Again, the necessary precautions are adequate forced-draught ventilation and atmospheric testing before entry and throughout the working cycle. Where there is a need to enter a dangerous or unknown atmosphere, appropriate breathing apparatus should be worn.

Hydrogen sulphide. Hydrogen sulphide in high concentrations can cause death, and at low levels, irritation to the eyes, nose and throat, dizziness and headache. It may generally be found in sewage, other decomposing organic materials and excavations in made-up ground. The necessary precautions are similar to those indicated for carbon dioxide. Workers carrying out maintenance in existing sewers should be particularly on the alert and should not enter manholes etc. without carrying out checks of the atmosphere, and should then undertake periodic checks while working underground.

Nitrous fumes. Nitrous fumes are very toxic: exposure to high concentrations may prove fatal and exposure to lower concentrations may result in bronchitis. A common source results from the use of explosives in tunnelling and excavating, for example. The confined space of the tunnel and excavation is perhaps the area of greatest risk. The main precaution is, if possible, to ensure that personnel are removed from the danger area during blasting and firing operations and that they stay away until the fumes have dissipated.

Cadmium. Breathing cadmium fumes would almost certainly result in cadmium poisoning, which occurs suddenly, causing drowsiness, vomiting, loss of muscular control and possibly death. Cadmium fumes are released during cutting, welding and brazing operations with cadmium-plated steel.

The main danger occurs when the operations are carried out in confined spaces such as tanks and fabricated structures, and the necessary precaution to take is the provision of good ventilation. It will often be necessary to provide local exhaust ventilation, especially in confined spaces, or to ensure the use of air-line breathing apparatus.

Lead. Lead poisoning may arise from breathing either lead dust or lead fumes and results in fatigue, anaemia, colic or wrist-drop. The cutting and burning of old structures covered in lead-based paint releases lead dust and fumes and therefore the highest risk area is probably demolition work.

Where possible the process should be segregated and appropriate (possibly local exhaust) ventilation provided. Good protective clothing should be worn (i.e. a type that does not hold dust), and in some cases respiratory protective equipment should be used.

Overalls or other protective clothing should be removed before meals are taken and not stored or taken into areas where food and drink are consumed. It is also important that users should wash immediately following the removal of all protective clothing. Similar standards of personal hygiene should be followed by operatives before they eat, drink or smoke.

Welding fumes. In welding operations the parent metal and its coatings, together with the weld metal and rod coating, release a variety of complex fumes which, if breathed, may give rise to metal-fume fever – a flu-like illness. The necessary preventive measure is the provision of local exhaust ventilation and, for short periods, respiratory protective equipment of the type that supplies fresh air. Moreover, as with lead, good personal hygiene is very important.

Zinc. Zinc fumes are evolved from the welding, brazing and flame cutting of galvanised steel, and breathing them may cause zinc-fume fever. Exhaust ventilation and respiratory protection are the precautions to be taken if this hazard is to be avoided.

Solvents. Various types of solvent are contained in a wide range of products used in the construction industry, e.g. adhesives, sealers, paints, lubricants and lacquers. Not only may they cause dermatitis through skin contact but they may also be a source of vapours which, when inhaled, can cause a variety of illnesses including headaches, dizziness and vomiting, and, in severe cases, unconsciousness preceding death. They are particularly hazardous when used in confined spaces (e.g. basements) because of higher concentrations and poor natural ventilation. A good standard of ventilation and appropriate protective clothing should be adopted to minimise these hazards.

Ingestion of harmful chemicals

The swallowing of harmful chemicals by mistake or through lack of personal hygiene is an ever-present risk in construction and demolition activities. The hazards and precautions against the ingestion of harmful chemicals are largely covered by the commentary above on dangers arising from contact with the skin and inhalation of harmful chemicals. The promotion of good personal hygiene, the provision of correct protective clothing, equipment, toilet and washing facilities, accompanied by training and supervision, are also essential to prevent the transmission of chemicals to the mouth from hands, arms, gloves, etc. In addition, there is a need for clear correct labelling of harmful substances with special attention paid to liquids not in their original containers, to warn against accidental swallowing.

Physical hazards

In the second main category of hazard danger results from the general environment experienced by the workers or from their particular occupation.

Cold

Construction workers may be exposed to conditions of extreme cold and to biting winds accompanied by rain over long periods. Such prolonged exposure without adequate protection may have serious effects on their health. The effects may range from a dulling of mental faculties and slower muscle reaction to possible bronchitis and arthritis. Furthermore, errors and accidents are likely to increase when the hands of operatives are particularly cold.

The first line of defence against the cold is to wear suitable protective clothing, supplemented by sensible undergarments and warm footwear. Moreover, if possible, overall general protection to the work area should be provided, e.g. the use of reinforced plastic sheeting in the form of screens. The provision of warm accommodation where hot drinks and food are available is essential.

Heat

Although not as prevalent in Britain as in countries with much hotter climatic conditions, heat stress in construction workers should not be totally discounted. The likely occurrence depends on the ambient air temperature, the relative humidity, radiant heat and the amount of air movement. Thus hot humid weather, especially for people carrying out heavy manual work in confined spaces with little air movement, provides the conditions for this hazard.

Another area may be underground, during tunnelling operations where heavy manual work is being undertaken in the vicinity of heavy machinery, and where the ventilation standard is not as good as it should be. The symptoms are headache, giddiness, fainting or muscular cramps, and victims should be taken to a cool place, provided with fluid to drink and allowed to rest. Where the case is severe and the casualty appears to be ill, develops a high temperature and does not perspire, urgent medical attention is necessary.

Noise

Damage to hearing, usually irreversible, may be caused by exposure to high noise levels from plant and machinery on site or in a workshop. Sound levels are measured by meters that indicate the sound pressure levels with a weighting factor to imitate the characteristics of the human ear, quoted as dB(A). The generally acceptable sound level upper limit is 90 dB(A) where the sound is reasonably steady and exposure is continuous for eight hours in any one day. However, many would regard 85 dB(A) as the upper limit and employers have to take certain basic steps at this level. Where the sound level fluctuates or the exposure period is other than eight hours, the daily personal noise exposure may be calculated by the use of formulae given in the HSE guidance on the Noise at Work Regulations 1989 *Noise at work.*[17]

In addition to the continuous exposure limits there is a requirement that the unprotected ear should not be exposed to a peak sound pressure level exceeding 200 Pa. Noise reduction should in the first instance be made by engineering control, e.g. more efficient silencers to the exhaust systems of air-cooled diesel or petrol-engined plant and pneumatic-powered equipment, sound-reducing covers for compressors, careful and regular maintenance of plant and machinery with particular attention given to the prevention of unnecessary vibration, and the use of acoustic screening of noisy fixed equipment where possible. Ear protectors should be a last resort and when they are used they must be carefully selected to suit the noise levels expected. Employers must provide protectors to all workers likely to be exposed to 90 dB(A) or 200 Pa and ensure that they are used – employees must wear them. Furthermore, protectors must be provided to employees who are exposed to between 85 and 90 dB(A) if they ask for them.

Vibration

A most common injury is caused by continued exposure of the hands to high frequencies of vibration from tools such as pneumatic hammers, concrete breakers, drills and chipping hammers. It is usually known as vibration white finger. It starts with a slight tingling or numbness in the fingers and eventually causes whiteness to the tips. Attacks may last about an hour and end with a sudden rush of blood to the affected tips, often causing considerable pain.

This occurrence can be reduced by the use of vibration isolators, by the operator exerting the lightest pressure on the tool commensurate with proper control and the provision of breaks from the work on to non-vibration work for the operator. The maintenance of good circulation and warm hands is also helpful and the wearing of comfortable gloves can be an asset.

Ionising radiation

Sealed radioactive sources are widely used in the industry to check welded joints in pipelines, for example, and for other non-destructive testing. Experience has shown that site radiographers are more likely to receive serious overdoses of radiation than are their factory-based counterparts. Those working in nuclear power stations, processing plants and laboratories may also be exposed to ionising radiation. The effects of exposure to ionising radiation may be radiological dermatitis, skin burns, loss of hair and bone cancer. Compliance with the Ionising Radiations Regulations 1985 together with the HSE's Approved code of practice *The protection of persons against ionising radiations arising from any work activity*[18] is, of course, essential, and this should obviate the likelihood of persons receiving excessive exposure.

In brief, only classified persons are allowed to undertake site radiography and their work must be carried out under the supervision of an appointed radiation protection adviser who should ensure that the work is carried out in accordance with the requirements of the regulations and that any local rules relevant to the work are observed.

Other workers (not involved with the radiography) and even the public may be at risk from scattered radiation. This may be reduced by the work being carried out within a suitable enclosure from which all but the classified workers are excluded. Where this is not reasonably practicable a designated area must be established, the boundary of which should be a safe distance from the exposed isotope. The boundary must be clearly indicated with warning signs or lamps at appropriate intervals and consideration should be given to posting sentries along the boundary as an extra precaution to ensure that no unauthorised person enters the danger area.

Compressed air

Working in air at pressures above atmospheric pressure, for example on diving activities, tunnelling or caisson construction may result in compressed-air illness. In its mildest form (type I) this may take the form of skin irritation, slight pains in the joints and tightness of the chest, but with the more serious form (type II) severe pains develop in the joints, and dizziness, unconsciousness and even death may occur. A particularly distressing type of compressed-air illness is aseptic bone necrosis, which affects the ends of the long bones in the body at the shoulder, hip and knee joints, for example. It leaves the affected person with symptoms similar to arthritis, varying from extreme discomfort to total disablement.

The cause of the illness is associated with too-rapid decompression and, to reduce the likelihood of its occurrence, compliance with acceptable decompression procedure and tables is necessary. The Work in Compressed Air Special Regulations 1958 include decompression tables for use in tunnelling and caisson work, but subsequent research has identified a decompression procedure which gives better protection than the 1958 tables, as they are known. This is incorporated in decompression tables known as the 1966 Blackpool tables and the Health and Safety Executive advises their use. Since the 1958 tables are the statutory requirement, contractors wishing to use the Blackpool tables must first obtain official approval from the Health and Safety Executive. Decompression procedures in respect of diving operations are covered by the Diving Operations at Work Regulations 1981, amended 1990 and 1991 (*see* Appendix 4).

Lasers

Lasers are used in the construction industry as an aid to setting out works which involve straight lines as, for example, in tunnels, runways and dredging work.

Eye injury, which could be severe, may occur if someone looks directly into the beam of a laser in the direction of the source. Moreover, contact may cause skin burns. It is important to select safe instruments and where possible to ensure the separation of personnel and the beam by means of effective barriers; furthermore, notices should be displayed to warn people against looking down the beam. Where people need to approach the beam, appropriate eye protection against the particular type of laser must be provided.

Manual handling injury

There is a further form of physical injury and that is the effect of sprains and strains caused by over exertion when manual handling. The Manual Handling Operations Regulations 1992 came into force in January 1993 and implement the EC Directive 90/269/EEC concerned with the 'minimum health and safety requirements for the manual handling of loads where there is a risk particularly of back injury to workers'. The regulations apply across all industries including the construction industry and place duties on employers, employees and the self-employed.

Employees have a duty under the regulations to follow the system of work 'laid down by their employers to promote safety during the handling of loads'. The self-employed are required to take the same steps to protect themselves from injury as employers need to take to protect their employees in similar circumstances.

The Health and Safety Executive has produced supporting guidance to the regulations (see Appendix 4). Its introduction points out that more than a quarter of all reported injuries each year are associated with manual handling. Further it draws to the reader's attention that many manual handling injuries are cumulative rather than being attributable to any single handling incident.

The regulations define 'manual handling operations' as any transporting or supporting of a load (including lifting, putting down, pushing, pulling, carrying or moving) by hand or by bodily force. The human effort may be applied directly to the load or indirectly by, for example, hauling on a rope or by pulling a lever.

The first duty of an employer is, so far as is reasonably practicable, to avoid the need for employees to undertake any manual handling operation which involves a risk of their being injured. The employer must therefore identify by a risk assessment the tasks that may be a safety risk. Such a risk assessment is required by the Management of Health and Safety at Work Regulations 1992 (see p. 194).

Having identified the manual handling operations that involve un-acceptable risk of injury the employer must try to identify a different method of work for example by the use of a mechanical aid. Where it is not reasonably practicable to obviate the need to handle the loads manually, an assessment of the operation must be made taking into account the questions posed in Schedule 1 of the regulations including

- the tasks – do they involve twisting the trunk?
- the loads – are they difficult to grasp?
- individual capability – does the job require unusual strength, height etc.?

This assessment may then form the basis for devising practical solutions to the handling problems. Having carried out the assessment, the employer must take the appropriate steps to reduce the risk of injury to the lowest practicable level. Employers must also provide their employees with information about the weight of each load and of the heaviest side if the centre of gravity is off centre.

Examples of practical solutions include

- team handling, i.e. using two or more persons (Fig. 4.45)
- mechanical aid such as a trolley or bogie that requires manual loading and unloading
- hydraulic hoists — several types are available
- breaking down the load into manageable components
- using personnel strong enough for the task to be undertaken safely.

If the employer believes that the assessment of manual handling oper-ations is no longer valid, for example if an injury occurs, or if there has been a significant change in the circumstances, a re-assessment must be carried out and any necessary changes to the method of operations made.

Fig. 4.45. Team handling

Biological hazards

Biological hazards experienced in the civil engineering and construction industry are likely only to affect those engaged on waste land sites or sewer outfalls or in sewers contaminated by vermin. The disease that is sometimes caught by these workers is leptospiral jaundice or Weil's disease. This is a feverish condition caused by ingestion of food or water contaminated with the urine of rats. Where not treated early and properly the disease can be fatal (Fig. 4.46). The contaminated sludge in sewers may also enter the skin or be absorbed through mucous membranes of the eyes, nose and mouth (*see also* p. 135).

 LEPTOSPIROSIS (WEIL'S DISEASE) CARD

If you work in or near sewage or in places where rats and other vermin are likely to live, you are at risk from various diseases. These can and must be treated.

Show this card to your doctor or hospital if you feel severe 'flu symptoms or if you have any sign of jaundice.

Further guidance can be found in the Company Health and Safety Procedures Manual at CSM/OHH/08 issued with this card

WS/Atkins

Most of the diseases connected with sewage enter the body through cuts and abrasions in the skin, including the mucus membrane lining the mouth and nose.

Prevention is by good hygiene and by ensuring that **all** cuts, however small, are covered with a waterproof plaster before work begins. In the event of an injury, ensure immediate first aid is supplied.

If you are frequently exposed to contamination, ask your doctor to record the fact on your notes.

Signature

WS/Atkins

Fig. 4.46. Typical Weil's Disease card (front and reverse)

5 Safety policies

Health and safety policies

Legislation

Section 2(3) of the HSW Act states

> Except in such cases as may be prescribed, it shall be the duty of every employer
> to prepare and as often as may be appropriate revise a written statement of his
> general policy with respect to the health and safety at work of his employees
> and the organisation and arrangements for the time being in force for carrying
> out that policy, and to bring the statement and any revision of it to the notice of
> all his employees.

The opening phrase is covered by the Employers Health and Safety Policy
Statements (Exemption) Regulations 1975, which exempt from these requirements 'any employer who carries on an undertaking in which for the
time being he employs less than 5 employees'. For the purposes of this
exemption, all the employees of an undertaking count whether they are
employed in one or several sites or establishments.

Put another way, the HSW Act demands that employers of five persons
or more

- state in writing their general policy on health and safety
- describe the organisation and arrangements for carrying out that
  policy
- revise the statement whenever appropriate
- bring the statement to the notice of all employees.

There are a number of regulations made under the HSW Act that place
specific duties upon employers and employees. They include, in the
context of this chapter, the Management of Health and Safety at Work
Regulations 1992 (*see* p. 194), and the CDM Regulations (*see* p. 20). These
must be taken into account when preparing safety policies and when
preparing terms of reference for managers and other employees with
responsibility.

Definition of 'employer' and 'client'.

Professional staff should not be confused by the use of the word 'employer' in the HSW Act and 'Employer' in the ICE Conditions of Contract. The former means 'one who employs persons in return for a wage or salary' while the latter is essentially the promoter of civil engineering works otherwise known as the client. Under the HSW Act, employers include all authorities, undertakings and companies, consulting engineers, architects, contractors, subcontractors and anyone else who has people under a contract of employment. The CDM Regulations use the term 'client' to mean the person for whom the project is carried out.

Health and safety policy statements

Advice to employers

Policy statements prepared and issued by employers to comply with the HSW Act should be in terms that can be clearly understood by their employees. The Health and Safety Commission advises employers to emphasise the legal significance of the statement by referring to the HSW Act in the opening paragraphs, and the notice should make clear not only the intentions of the employer but also the obligations of employees. Policy statements should indicate how the company is organised with respect to the health and safety responsibilities of management and indicate its commitment to providing information, training and advice to employees.

Any arrangements the company has made for the establishment of safety committees and for consultation with safety representatives as required under section 2(4), (6) and (7) of the HSW Act should be covered in the policy statement. Finally, the statement should be signed by the chief executive, the managing director or other responsible head to make clear the commitment to the policy at the very highest level.

Policy statements applied to construction work

The framing of the policy will, of course, be influenced by the type of undertaking or business conducted by the company. Most employers engaged in promoting, designing, constructing and demolishing civil engineering works and buildings have responsibilities, both to their employees, and to other people, who may be employees of other companies associated with the project, visitors or members of the general public.

The opportunity should not be overlooked to emphasise to all employees the need to act at work in such a way that the health and safety of the public and of people who work for other organisations are also protected. A health and safety policy statement should therefore contain the following basic elements.

- That the company has a legal duty, as far as is reasonably practicable, to provide safe and healthy working conditions for its employees.
- That the company will conduct its work in such a manner that the health and safety of others, besides that of its employees, is not adversely affected.
- That employees are required to co-operate with the company in preserving their own health and that of other employees and anyone who might be affected by their activities at work.
- That an organisation and chain of authority in matters of safety has been set up.
- That safety information, instruction, training and supervision will be provided for all employees as and when necessary.
- That the company will encourage the work of a safety committee and co-operate with safety representatives if that is the employees' wish.

Typical policy statements

Health and safety policy statements for organisations involved with construction have many similarities. Differences lie in the length and method of presentation. Some employers choose to display the statement on notice boards while others include it in a safety handbook. Either method is acceptable provided the message reaches all employees.

One consideration is the problem of the changing organisation, particularly in a large company. Because the organisation for implementing the health and safety policy should contain the names of staff with posts of responsibility, it is usual to publish an organisation chart as an appendix to the main policy statement to avoid the need to reprint and reissue it every time there is a managerial change. Some organisations publish their policy statements in a loose-leaf binder for easier updating.

Policy statement of a major employer

It is convenient to describe the health and safety policy of a major employer by reference to a fictitious 'Railway Company'. This railway employs a large number of personnel on maintenance duties. New construction is carried out sometimes by direct labour and sometimes by independent contractors. The Railway Company is at times client, designer and contractor. Apart from occupational health and safety, the safety of the travelling public is of paramount importance.

The Chairman of the Board carries the primary responsibility for health and safety on the Railway. He issues a personal policy statement which emphasises the Board's legal duty and its commitment to health and safety and welfare of its employees and of those who may be affected by their activities. This is displayed on all staff notice boards. He also issues instructions to each of his directors, one being the Director of Civil Engineering, to prepare and implement detailed safety policies in each of the operating regions of the railway. These are known as local safety policy statements and they are issued to staff from supervisor grade upwards. It is normal practice for the General Manager of the region to sign the regional engineer's policy statement, ensuring an essential link in cascading the policies to supervisors.

A typical local policy statement includes details regarding

- responsibility: the names of staff responsible for safety, together with their locations and delegated duties
- safety representative: arrangements for committees and trade union consultation on safety matters
- inspections: management inspections to identify risks arising out of unsatisfactory workplaces, working systems or behaviour
- induction: safety instruction for new employees
- training: arrangements for further training
- accident reporting: internal arrangements for reporting and recording personal injuries and for accident investigation
- plant: maintenance and operator training
- materials: approval of substances and materials that may be of a hazardous nature
- routine processes: establishment of safe working systems
- protective clothing: arrangements for selection and wearing of protective clothing and safety equipment
- fire/explosion: reference to a separate fire manual and appointment of fire wardens
- welfare: assistance with staff welfare problems
- first aid: provision of first-aid equipment and first-aiders
- procedures for complying with CDM Regulations
- contractors and other non-railway personnel: reference to a separate booklet entitled *Track safety handbook* which must be supplied to contractors and other railway staff responsible for contractors' safety
- safety advisory service: how to seek advice.

The statement concludes with a number of appendices giving local information which must be completed by the local manager concerned, and lists of employees who should receive the various other railway safety publications and booklets.

The general safety policy of the Railway Company is typical of most major employers. However, the local or particular policy statement, which reflects more closely the special hazards of the industry or the service it provides, varies in content depending on the emphasis needed. Nevertheless, policy statements are expected to outline the measures by which the management safeguards its employees, contractors and other people on the premises and ensures the safety of the public, who could be at risk if the procedures laid down are not complied with.

Policy statement of a contractor. A typical safety policy statement of a building or civil engineering contractor is suggested here. The fictitious contractor, Constructwell Limited, includes its policy statement in the company safety booklet.

To all Employees of Constructwell Limited
Health and Safety Policy

You will know from your Contract of Employment that you are an employee of Constructwell Limited. Under the Health and Safety at Work etc. Act 1974 the Company is required to have a Health and Safety Policy and to bring it to your attention. You are required under the same Act to follow the safety procedures adopted by the Company.

This notice sets out the Company's Health and Safety Policy and the organisation and arrangements adopted for its implementation. It sets down the procedures which management will apply in the interests of health and safety of all employees.

I wish, however, to emphasise that each and every one of us has a duty to take reasonable care of ourselves and of those who work with us. We must all work together to prevent accidents and the hardships that follow.

Signed . . . . Issued . . . . 19 . . . .
Chairman

General Policy Statement
It is the policy of Constructwell Limited

(a) to ensure the health and safety of all its employees while at work by providing, so far as is reasonably practicable, working environments that are safe and without risk to health

(b) to conduct the undertakings in such a way as to ensure, so far as is reasonably practicable, that people not in its employment, but those who may be affected, are not exposed to a risk to their health

(c) to recognise its obligation to meet all relevant legislative requirements pertaining to health, safety and welfare which apply to any of the Company's undertakings

(d) to organise and arrange its affairs to ensure effective implementation of the policy.

Operation of the Policy
In carrying out its general policy it is the practice of Constructwell Limited

(a) to specify in writing managerial responsibility and accountability for the safety, health and welfare of its employees and for the health and safety of others who may be affected by its undertakings and to bring this to the attention of all employees

(b) to ensure that appropriate safety training and instruction is provided for all employees and that accident prevention is included in all relevant training programmes, particularly those for graduates, trainees, apprentices and other young employees

(c) by a programme of regular presentations and publicity to sustain an awareness of the need to prevent accidents and health risks in the minds of all its employees

(d) to take into account when planning its work any aspects which help to eliminate risk of injury or health damage to employees or others

(e) to require subcontractors to operate a safety policy no less stringent than that of the Company; to provide safety training for subcontractors' employees and self-employed as necessary

(f) to make appropriate accident prevention arrangements by effective liaison with employers other than subcontractors who have employees working at the same work place as the Company

(g) to encourage the discussion of health and safety matters at all levels including the setting up of arrangements for joint consultations with employees through their appointed safety representatives.

Organisation

Managers at all levels are responsible for the health and safety at work of employees reporting to them and others to whom the company has a duty. They will ensure that safe working conditions and welfare facilities are provided in accordance with statutory regulations and company policy and, so far as is reasonably practicable, carry out works in such a manner that there is no risk to other people.

The line management structure of Constructwell Limited is published from time to time and displayed on principal notice boards.

The Safety Service

A senior corporate director will be responsible for ensuring that an effective safety service is provided. He will appoint a company safety service manager and a number of suitably qualified safety advisers.

The main responsibilities of the safety service are

(a) to ensure that managers understand their duties as outlined above and to assist and advise them in the execution of such duties

(b) to inspect and report on places of work to ensure that safe systems are being employed in compliance with statutory regulations and company practices

(c) to warn managers and other employees of existing and potential health hazards

(d) to investigate accidents, analyse accident data, report and make recommendations to prevent recurrences

(e) to ensure that all reports and records required by statutory provisions are made and maintained

(f) to promote safety training for all employees and, where necessary, subcontractors' employees and self-employed.

Constructwell Limited is a sufficiently large company to warrant the establishment of a safety service with a number of safety officers. Smaller contractors have to organise their safety arrangements in other ways, as suggested later in this chapter, although in principle the general policy on health and safety would be common to most civil engineering contractors.

Policy statement of a consulting engineer or architect. The following is a suggested safety policy statement for a firm of consulting engineers or an architectural practice.

Victoria Street & Partners
Health and Safety Policy Statement

The Health and Safety at Work etc. Act 1974 (the Act) places a duty on every employer to prepare and, as often as may be appropriate, revise, a written statement of his general policy with respect to the health and safety at work of his employees and the organisation and arrangements for carrying out that policy, and to bring the statement to the notice of all his employees.

The Company Policy

It is Victoria Street & Partners Policy in so far as is reasonably practicable

1. to provide and maintain places of work for their employees in conditions that are safe and without risk to health together with adequate facilities for first aid and welfare
2. to conduct their business in an organised and responsible manner and to adopt systems of work that preserve the health of their employees and other people concerned in business activities
3. to provide lines of communication to enable employees to co-operate in promoting and developing effective measures to ensure their health and safety at work
4. to give information, instruction, training and supervision as necessary to ensure that their employees are aware of their responsibilities under the Act
5. to make available personal protective clothing and equipment for employees whose occupation or place of work is likely to be hazardous.

The Employees' Responsibilities

The Act requires that employees while at work

1. shall take reasonable care of their health and safety and that of all other persons who may be affected by their acts or omissions at work
2. shall co-operate with their employer in all safety, health and welfare matters and maintain their place of work and all their equipment in a tidy and safe condition
3. shall not interfere with or misuse anything provided in the interests of their health and welfare.

Further details of the employees' responsibilities in connection with the Act are given in the Company Safety Handbook.

The Health and Safety Organisation

The means by which Victoria Street & Partners safety policy is implemented is through line and project management. Annexe A outlines the line-reporting relationships within senior management with respect to health and safety.

The responsibility for safe systems of work on projects and for the health and safety of employees working on projects away from the company premises rests with project managers, one of whom is assigned to every project by the appropriate partner in charge.

Staff with Special Responsibilities for Safety

The partner of Victoria Street & Partners nominated in Annexe A is accountable to the senior partner for the health and safety policy. He has the direct staff assistance of

1. the Company Safety Adviser, who is responsible for advising the partners of their duties under the Act in relation to
    (a) the health and safety of employees while engaged in business duties while outside the company premises
    (b) the organisation of safe systems of work so as to protect other people from risks arising out of work activities
    (c) training and informing employees about their duties under the Act, the hazards of their work and precautions to be taken;
2. the Administration Manager, who is responsible for the health, safety and welfare of employees while on company premises.

Display of this Policy Statement
The latest revision of this statement, together with the organisation chart Annexe A, will be displayed on principal notice boards by the Administration Manager.

Signed. . . . . Issued. . . . . 19 . . . .
Senior Partner

For every project Victoria Street & Partners appoint a project manager, who is responsible for all aspects of the project to which he is assigned. His duties will be outlined within his terms of reference. Other firms of consultants prefer to use the title 'project engineer'. Whatever title is adopted, it is essential that one person of suitable experience and seniority is responsible for each project and that the reporting relationships on that project are clearly understood by all staff working on it. On large projects the duties of the project manager may have to be delegated, and it is then necessary to give clear terms of reference to all subproject managers if the health and safety objectives are to be achieved.

It must be assumed that Victoria Street & Partners is a large firm situated in one principal office. Smaller companies may well combine the health and safety responsibilities of the company safety adviser with that of the administration manager. Other firms have several offices and the duties of the two people in our fictitious consultancy must be covered in other ways. Whatever managerial arrangement is preferred, the company safety adviser must be competent in health and safety matters to comply with the Management of Health and Safety at Work Regulations 1992. The organisation diagram should clarify these responsibilities to the employees. The employees of Victoria Street & Partners do not have a safety committee, otherwise there would be some reference to this in the policy statement. Where employees wish to have safety committees and safety representatives, employers must co-operate with them.

Organisation

The type of business

The organisation by which a company fulfils the aims of its health and safety policy reflects its particular business purpose. Whereas contractors are concerned almost entirely with the safety of their employees and others on construction sites, the major employing authorities (e.g. the electricity, gas and water companies, airports and railways) have a special responsibility for the protection of the public when construction work on their premises or involving their apparatus is being undertaken. As a result their safety organisation must be set up to cater for the risks associated with their type of business.

Companies offering professional services such as consulting engineers, architects and quantity surveyors have a legal duty to see that the designs, specifications and other documents issued for construction do no harm to the health or safety of anyone working on the project or affected by it. Therefore their safety organisation must establish effective managerial control over the health and safety aspects of the designs and their checking. They also have employees whose safety must be considered.

All firms involved in construction activities whether they be clients, contractors or professional advisers, have duties under the CDM Regulations and their safety policies and organisations have to establish how these duties are to be implemented by managers and staff within the company framework.

Duties of management

The chief executive, managing director or senior partner must assume ultimate responsibility for the safety policy and normally sets up an organisation based on the line management structure to assist in fulfilling this responsibility. Figures 5.1–5.4 give examples of the reporting lines of staff with managerial responsibilities in typical authorities, contractors and consulting engineers.

It is common practice in organisations of more than about 50 employees for the chief executive to appoint a senior colleague as safety director to take special responsibility for safety on the management board. While the appointment of safety director is usually combined with other duties, the job is to set up and maintain an organisation to promulgate the safety policy and report to the board or management committee on all safety matters. In smaller companies the chief executive may carry out these functions.

The management committee must make adequate financial provision for safety, health and welfare of employees, and one of the duties of the safety director is to manage the budget allocated to maintain the safety organisation and provide information and training. Financial provision for safety measures must also be allowed in tenders for design or construction work. Estimators and project managers must be made aware of this. The duties of the safety director or the senior manager responsible to the chief executive for safety matters may be summed up in typical terms of reference which are given later.

The diverse nature of the construction industry puts considerable onus for safety awareness on the managers of departments, subsidiary offices, project offices and sites. It is probably true to say that the success or failure of a safety policy rests with this level of management. One of the important tasks of a safety director is to ensure that all construction and site managers and other senior staff are totally committed to the company safety policy. Terms of reference or standing instructions must be provided for senior staff and the safety director will need to meet senior staff from time to time to ensure that their duties are understood and carried out.

Safety organisation of a major employer

Figure 5.1 shows the structure of responsibility for health and safety within the civil engineering departments of a fictitious 'Railway Company'. The responsibility for implementing the Board's safety policy rests with line

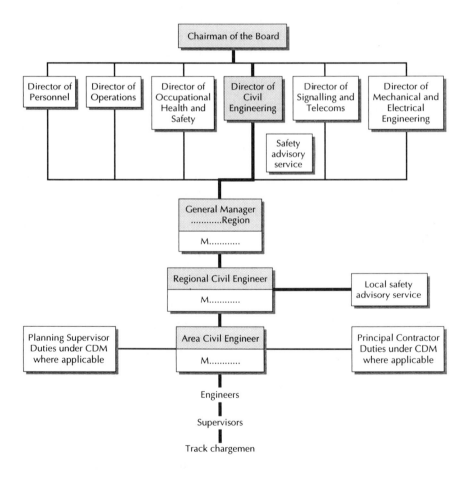

Fig. 5.1. A major employer – Railway Company: structure of responsibility for health and safety in civil engineering

management from the track chargeman up to the Chairman. The safety departments set up by the Director of Civil Engineering and the Regional Civil Engineers are clearly labelled advisory to emphasise this point.

Copies of the full local safety policy statement are given to all staff down to engineer level and, in a modified form, to supervisor level. All policy statements leave blanks for the names of staff to be inserted, leaving no doubt about either the identity of the person who is to carry out the actions or the identity of the person to whom he or she reports. The safety departments themselves have organisational structures and reporting procedures to suit the measure of hazard that the particular undertaking poses to workforce or public.

Major employers carrying out construction work whether by outside contractors or by in-house works units become clients under the CDM Regulations. For projects that are likely to come within the regulations (*see* p. 20) the appointment of designers and contractors and the appointment of a planning supervisor and a principal contractor carry with them the duty to see that those appointed are competent.

The work must not start until the client is satisfied that the principal contractor has prepared a sufficiently advanced construction stage health and safety plan. The principal contractor must then develop this plan in accordance with the requirements of CDM Regulation 15 (4). The planning supervisor must ensure that a health and safety file is completed and kept available by the client for inspection at any time in the future.

An organisation must be set up to carry out these tasks, and the terms of reference for managers involved with construction must state clearly who are to carry out the duties necessary under the CDM Regulations, who checks that they are done and where the safety file is to be kept after completion.

Safety organisation of a contractor

For a contractor the primary objective of the safety organisation is the prevention of accidents on construction sites and the diagram has to show the line management responsible for achieving this. Figure 5.2 shows the structure of responsibility for health and safety in a major building and civil engineering contracting company. In this case, all the directors are responsible to the chief executive for the safety of the sites under their jurisdiction and that responsibility is carried down through, but not delegated to, the construction managers and site managers.

One of the directors has special responsibility for providing the safety service as others have responsibilities for the estimators, design office and plant departments. All these service departments contribute to the corporate safety policy in roles that essentially support the construction and site managers.

Terms of reference for the safety director. Typical terms of reference for a contractor's safety director are as follows

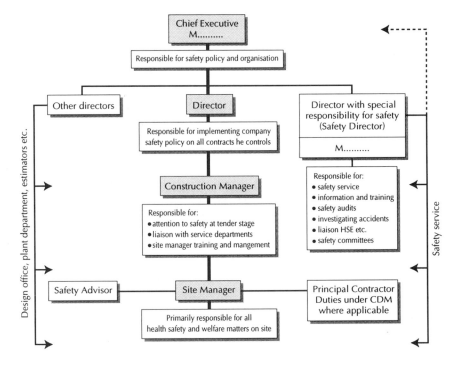

Fig. 5.2. Civil engineering contractor: company structure of responsibility for health and safety

To ensure so far as is reasonably practicable that the company meets its legal obligations with respect to health, safety and welfare of its employees and of others who may be affected by its acts or omissions. All are set out in the company health and safety policy statement.

In particular the safety director must

(a) administer a safety organisation, appoint appropriate staff and provide them with terms of reference

(b) ensure appropriate information, instruction and training for all staff and operatives

(c) provide facilities for first aid and welfare

(d) prepare annual budget estimates for safety, health and welfare facilities and manage the allocation of funds accordingly

(e) be responsible for reporting injuries, diseases and dangerous occurrences and keep the Board regularly informed of such events

(f) ensure adequate provision of time and costs of safety measures in all design, construction and related activities

(g) ensure that any safety committees proposed by staff or operatives are properly constituted, attended by management at the appropriate level, accurate records of meetings are made and kept, and that the resolutions put forward by safety representatives are acted on to the committees' satisfaction.

Terms of reference for a construction manager. The terms of reference for a construction manager responsible for a number of sites each under the control of a site manager are primarily to see that company policy is implemented on those sites. Construction managers must ensure that adequate consideration has been given to site health and safety at tender stage and that the safety department, plant department, design office and any other support facilities are made available to the site manager.

Construction managers must also see that the site managers have been thoroughly trained in their duties for health, safety and welfare matters and that the duties are carried out. Advice given by safety advisers must be implemented and, where advice is ignored by any site manager, the construction manager must take appropriate action. Through liaison with the company safety adviser, the construction manager will ensure that the site managers are given all the support they need.

Where the client has appointed the contractor as 'principal contractor' under the CDM Regulations, the construction manager has to allocate to one or more individuals their specific tasks and see that they are carried out. In many, indeed in most, cases it will be the site manager who performs the role of 'principal contractor' and the company safety adviser who checks that the duties are correctly performed.

Safety responsibilities of a site manager. Typical safety responsibilities of a site manager are

To organise the site in such a manner that it is safe both for company employees and other persons who may be affected.

In particular the site manager shall

1. comply with the HSW Act, the regulations concerning construction and other relevant legislation
2. appoint the necessary staff to carry responsibility for site safety and welfare and to provide them with the authority to act
3. complete the statutory notices and liaise with HM Factory Inspector and the chief officer of the local fire brigade and ambulance service
4. provide time and funds for safety measures
5. carry out written risk assessments and implement the necessary controls
6. co-operate with the 'principal contractor' and provide information required under the CDM Regulations including RIDDOR reports
7. ensure that all staff and operatives have the necessary training and follow correct working procedures
8. provide protective clothing and safety equipment where necessary and see that it is used
9. arrange for the company safety adviser to set up a site appraisal
10. ensure that the procedures to be adopted in the event of an accident are clearly understood
11. make arrangements to supervise the safety of subcontractors' work and self-employed staff and labour.

Where the company has been appointed 'principal contractor' the following additional duties are to be carried out

12. develop and implement the health and safety plan
13. ensure co-ordination and co-operation between all contractors on the site and see that they have information about special risks
14. obtain contractors' risk assessments
15. monitor contractors' safety performance
16. make sure only authorised people are allowed on site
17. pass information to the planning supervisor for the health and safety file
18. ensure that notification particulars under CDM Regulations are displayed in a readable condition in a position where they can be seen by anyone concerned with construction work on the project.

Safety responsibilities of the safety adviser or officer. Under the Management of Health and Safety at Work Regulations 1992 every employer must have one or more competent persons to assist him in complying with health and safety law and applying protective measures. Further, under the CDM Regulations the principal contractor has to ensure that the health and safety plan for the project contains arrangements for monitoring compliance with health and safety legislation. Also the regulations require all contractors to comply with any rules applicable to them in the health and safety plan.

In the construction industry the function of the safety adviser may be as a full-time head of department in a large company or authority or a part-time specialist with a small firm, and there are many variations between the two extremes. Some smaller contractors operate group schemes that share the services of a full-time safety professional.

Safety advisers employed by contractors should be thoroughly experienced in construction techniques and legislation and preferably qualified by membership of the Institution of Occupational Safety and Health. The safety adviser reports to the safety director, who provides the terms of reference. The adviser may well be assisted by a number of staff – indeed some larger firms are headed by a director of the safety service, with several safety advisers reporting to this director.

The position of safety adviser is essentially an advisory one. Hence the preferred title of safety adviser rather than safety officer. *No action of the safety adviser must diminish the clear authority of line managers in matters of safety.* Only in exceptional circumstances where health is at risk or there is imminent risk of an accident should a safety adviser exercise the authority given by the directors to override the processes of line management.

The function of a safety adviser with a firm of contractors is

- to advise management on its duties under health, safety and welfare legislation and any revisions to it
- to report on the fulfilment of the objectives of the company's safety policy
- to advise on safe working practices, safety of plant and equipment,

environment health problems, protective clothing and permit-to-work procedures
- to advise on the implementation of the CDM Regulations in so far as they apply to principal contractors and contractors.
- to prepare and circulate safety memoranda, guidance notes, posters and other printed information
- to organise safety training courses for staff and operatives
- to carry out site safety, health and welfare audits
- to take responsibility for the company accident record book, investigate and report on all accidents and recommend measures to prevent recurrence
- to advise on the special problem of subcontractors and self-employed operatives on the company sites
- to attend safety committees as required and give advice to managers in their consultations with safety representatives
- to act as the focus in the company's correspondence with the Health and Safety Executive and professional bodies specialising in safety and accident prevention.

Management contracting

Compared with ordinary contracting, management contracting, the process whereby the management contractor assigns the majority of the work to a number of works contractors, requires a different approach for the organisation of site safety. Whereas every works contractor is responsible for the safety of his own employees, it is clearly necessary for firms to co-ordinate their activities so that they act responsibly to one another and as an integral part of the total project. The need for co-ordination is emphasised by the fact that many subcontractors employed by the works contractors have fewer than five employees and are therefore exempt from preparing policy statements.

It is essential that the conditions of contract require the management contractor to take full responsibility for site safety, including complying with all the general site legal requirements and to co-ordinate the safety practices of all companies contributing to the project. In most cases the management contractor will be appointed 'principal contractor' under the CDM Regulations by the client. In this case, co-ordinating the safety conduct of all contractors on the site becomes a legal duty and of course there are several other tasks required of a principal contractor under the CDM Regulations (*see* p. 20).

Organisation of health and safety policy for management contracting. A typical statement of the organisation for implementing the health and safety policy of a management contractor is as follows. It will be noted that the Constructwell management contracting company is part of the Constructwell Group.

Constructwell Management Contracting Limited
Organisation of the Safety Policy

General Considerations

The Company recognises that the safety and health in all those affected by its undertakings is an inherent part of its total business and that all staff can make significant contributions to the prevention of accidents and ill health arising from its operations.

The health and safety policy of the Constructwell Group is applicable to its management contracting operations and the statement shall be brought to the attention of the staff and operatives of subcontractors in addition to Company staff.

The means by which the Company implements its health and safety policy is through line management. The primary responsibility for safety rests with the project managers, one of whom is appointed for every project. The project managers are to be supported in every respect by the directors and all other managers and staff. In addition, assistance is provided from the various service departments as follows.

Duties of Service Departments

The Constructwell Group Safety Service is available

1. for advice on all matters of health, safety and welfare
2. to carry out safety appraisals, audits and inspection
3. to investigate accidents
4. to promote safety training.

The Proposals Department must agree with clients, their professional advisers and with the planning supervisor the collective and individual responsibilities for site safety and in particular the arrangements for common services.

The Planning Department will

(a) ensure that tender and works programmes make due allowance for the safety aspects of the work and comply with the safety plan
(b) bring to the attention of the project manager and planning supervisor any areas of high risk for discussion with the Safety Service.

The Purchasing Department

(a) must ensure for CDM Regulations purposes that all contractors and sub-contractors are competent and that they are adequately resourced to carry out their work safely
(b) will ensure that contract documents encourage safe working and make proper provision for subsequent payment
(c) must study subcontract submissions to identify any factors that could lead to unsafe working
(d) must advise the project manager of any special or unusual requirements in the contract documents that might affect the safety of the workforce or others
(e) must ensure that provision for safety requirements is made in the project cost plan.

The Central Plant Engineering Department will

(*a*) advise on special lifts, installation, maintenance and inspection and use of cranes, hoists, lifting gear and all other lifting appliances

(*b*) supervise any plant items provided by subcontractors and advise the project manager of any non-compliance with relevant statutory requirements.

The Design Department will

(*a*) carry out the design of temporary works or check and approve the design by others as requested by the project manager

(*b*) carry out, or commission and approve the design of permanent works as requested by the project manager; due consideration is to be made of all aspects of safety and health matters in construction, operation, maintenance and demolition

(*c*) note that a quality plan is to be prepared and followed for every design project.

Duties of the project manager

The project manager will give careful consideration to all aspects of health and safety on site. He will ensure that work plans of each of the works contractors include detailed provision for safety and also ensure that any shortfall in the responsibilities of subcontractors is rectified by instructions in writing.

The project manager will assume the duties of 'principal contractor' under the CDM Regulations where instructed to do so. Otherwise he will co-operate with the principal contractor, provide all information required by the regulations and act upon instructions from the principal contractor in connection with his statutory duties.

Particular consideration must be given to common services on the site and the allocation of responsibility for their provision and maintenance in accordance with appropriate legislation, e.g.

(*a*) liaison with fire, ambulance and rescue services

(*b*) liaison with the Factory Inspectorate

(*c*) first-aid and nursing facilities

(*d*) fire precautions

(*e*) health and safety posters, notices, information sheets and safety training

(*f*) protective clothing and safety and rescue equipment

(*g*) site offices

(*h*) canteen and toilet facilities

(*i*) access to the works including scaffolding

(*j*) lifting appliances

(*k*) temporary lighting

(*l*) fuel and power supplies

(*m*) overhead and underground cables.

The project manager will arrange for the subject of site safety to be included on the agenda at every regular progress meeting and will ensure that any matters raised are dealt with. He will encourage the formation of a site safety committee comprising representatives from all subcontractors. The chair of such committees will be taken by the project manager or his nominee, a Company staff member. Where employees wish to elect site safety representatives the appropriate arrangements will be made in accordance with section 2(4), (5) and (6) of the HSW Act .

The project manager will seek the advice and assistance of the service departments on any safety and health matters outside his certain knowledge. On acceptance of the bid he will initiate a safety appraisal to ascertain any particular hazards in the construction or demolition procedures and ensure that all concerned are made aware of the risks and precautions to be taken.

Subcontracting

The practice of extensive subcontracting has been adopted in recent years in some parts of Britain. Whereas a normal civil engineering contract under ICE Conditions recognises that some of the works may be sublet with the written consent of the engineer, many contracts are now executed largely by subcontractors, the contractor having only a cadre of managerial staff and key operatives on site.

With regard to health and safety, this situation is akin to management contracting and the principles of demarcation and responsibility for site safety are the same. However, in practice the effective control of site safety practices is difficult to enforce when a number of small subcontractors, especially those with fewer than five employees, are engaged on one site.

Nevertheless, it is desirable that engineers take into account the safety policy and method statements of the subcontractors when giving written approval to their employment together with the contractor's proposals for maintaining overall effective control. In any case the principal contractor has a duty to see that all contractors and subcontractors on the site are competent, are adequately resourced and will carry out their obligations under the CDM Regulations.

Safety groups and group safety schemes

Group safety schemes are operated in a number of areas and are intended for firms whose business or turnover does not warrant the employment of a full-time safety adviser or safety consultant. To comply with the supervision requirements of the Management of Health and Safety at Work Regulations 1992 and the CDM Regulations 1994, firms forming the group each contribute to the services of a group safety supervisor whose duties are

- to advise member firms of the requirements of the HSW Act, the Management of Health and Safety at Work Regulations 1992, the CDM Regulations 1994 and other relevant legislation
- to advise member firms on the preparation, organisation and implementation of their safety policies
- to advise member firms on risk assessments and risk control
- to organise safety information and training

- to carry out safety audits, site inspection and reports
- to carry out accident investigation and advise on action to be taken in the event of an accident
- to liaise with HM Factory Inspectorate and local authorities on behalf of member firms and the group.

Safety supervisors have to be experienced in the work being carried out and be suitably qualified. This also applies to safety advisers under the Management of Health and Safety at Work Regulations 1992.

The structure of responsibility for health and safety for a small contractor in a group safety scheme is shown in Fig. 5.3. Details on the formation of group safety schemes and suggested financial arrangements may be obtained from the Building Employers Confederation in London. While this implies that group schemes are applicable only to builders and contractors this is not so. Since the introduction of legislation requiring all employers to have competent safety advice, the formation of safety groups among firms of consulting engineers or architects to carry out similar objectives but with a bias towards design and supervision is recommended.

Safety representatives and safety committees

Section 2(4) of the HSW Act provides for the appointment in prescribed cases by recommended trade unions of safety representatives from among the employees. In the construction industry this applies to members of one of the unions which is party to the Civil Engineering Construction Conciliation Board or, in building, the National Joint Council for the

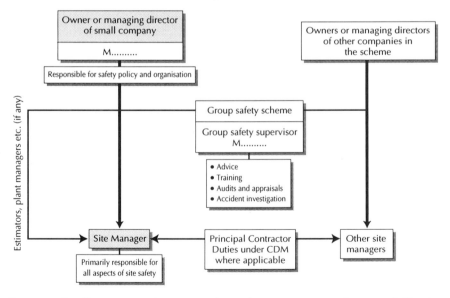

Fig. 5.3. Small contractor in group safety scheme: structure of responsibility for health and safety

Building Industry Administrative Council. Where requested by two or more safety representatives, employers have a duty to establish a safety committee with the function of keeping under review the measures taken to ensure the health and safety at work of employees and others.

The Safety Representatives and Safety Committees Regulations 1977 describe the appointment and functions of safety representatives and the establishment of safety committees. The Health and Safety Commission booklet *Safety representatives and safety committees* (*see* Appendix 4) sets out the regulations, the associated code of practice and guidance notes. The Commission makes it clear that, while employers have a duty to make arrangements with trade unions for the functioning of safety representatives and safety committees there is nothing to prevent employers and employees from making arrangements for joint consultation over health and safety at work that do not follow the provisions and advice contained in the booklet provided they are satisfactory to both sides.

The main functions of safety representatives are

- to investigate potential hazards and examine the causes of accidents at the workplace
- to investigate complaints by employees on matters of health and safety
- to make representations to the employer
- to represent the employees in consultation with inspectors of the Health and Safety Executive.

Under the strict requirements of the HSW Act, safety committees are set up as a result of trade union action to appoint the safety representatives who request the formation of the committee. However, the formation of *ad hoc* committees on construction sites where there are no elected safety representatives must be encouraged. Experience has shown that sites with good employer/employee consultation and good contractor/engineer relationships are also safe sites.

Safety organisation of a consulting engineer or architect

The duties of a consulting engineer or architect with regard to safety resemble those of the major employer and the contractor. On the one hand, professional staff, particularly when on assignment out of the office, are at risk of injury or health damage and, on the other hand, their business comprises the design and specification of items that might be hazardous to those who build, operate or maintain.

The safety of company staff is the responsibility of line management and the safety organisation chart should make this clear. A typical safety organisation for a firm of consulting engineers is given in Fig. 5.4. The project manager who carries the primary responsibility for the safety of staff on the project and for the quality of the design is supported by a safety adviser who acts in an advisory capacity. The considerable diversity of

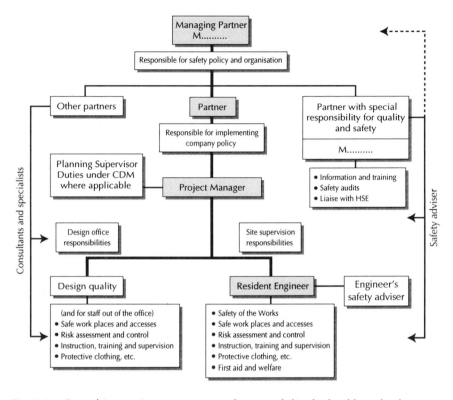

Fig. 5.4. Consulting engineers: structure of responsibility for health and safety

design work with which a consulting engineer is involved makes it impracticable for the safety adviser to advise effectively on all matters of design and operational safety. The project manager must therefore rely on the advice of consultants and specialists from within and outside the firm when called on to design items outside his common knowledge. An effective quality assurance policy will ensure that the safety aspects of designs and specifications have been properly considered.

For projects that are subject to CDM Regulations (*see* p. 20) a planning supervisor is appointed by the client. If the client decides to vest this appointment in a consulting engineer or architect, the firm of engineers or architects may decide to allocate the duties of planning supervisor to the project manager provided that the CDM criteria of competence and resourcing are met. The terms of reference for the project manager must make it quite clear what must be done to satisfy the CDM Regulations.

Terms of reference for the project manager. The terms of reference in regard to health and safety issued to project managers are to ensure that the project is designed and supervised in a manner that is safe for the company employees, contractors, personnel and members of the general public. In particular, the project manager must

- ensure that the client is aware of his duties under CDM Regulations
- see that the design department is adequately resourced by personnel competent in ensuring all aspects of health and safety are considered in the design phase of the project
- organise the design to minimise risk of error and ensure that the project can be safely built, operated and maintained
- consider methods that will protect workers if risks cannot be effectively overcome
- ensure that the design and drawings contain adequate information on health and safety and give this information to the planning supervisor to include in the health and safety plan
- ensure co-operation with the planning supervisor and any other designers
- ensure that company employees when out of the office on business have safe workplaces and means of access; organise written risk assessments for these activities and means of risk control as required; ensure these staff are aware of hazards, have received appropriate instruction and training, that they are adequately supervised and that they have the necessary protective clothing and equipment
- understand and comply with first-aid and accident reporting procedures in respect of employees out of the office on business duties
- in the case of site supervision, ensure that resident engineers have suitable terms of reference and are properly trained in health and safety matters and that the requirements of the design and the safety plan are clearly understood.

Where the project manager has been appointed planning supervisor he must in addition

- notify the project to the Health and Safety Executive if the project is notifiable under the CDM Regulations
- prepare the health and safety plan for inclusion with the tender documents
- give advice to the client on such matters as competence and resources of designers and contractors
- keep the health and safety file and hand it to the client on completion of the project.

Terms of reference for the safety adviser. Typical terms of reference for the consulting engineer's or architect's safety adviser are as follows.

The company safety adviser reports to the partner responsible for company safety. His main function is to advise the management board on their responsibilities under the HSW Act as described in the company policy statement. He co-ordinates with the head of administration who is responsible for the health, safety and welfare of staff in the premises occupied by the company.
   In particular he must

(a) arrange for the issue of booklets, guidance notes, memoranda, posters and literature necessary to inform staff of their duties under the Act

(b) initiate and propagate the required training in health and safety

(c) advise project managers about potentially hazardous situations likely to be encountered by staff and any other safety and health matters associated with the project

(d) give advice to project managers about policy on written risk assessments, control measures and the issue and use of personal protective equipment

(e) advise project managers of their duties under the CDM Regulations and company policy on managerial aspects of the planning supervisors' duties

(f) in an emergency instruct the cessation of any hazardous action by a member of the company

(g) act as the focus in the company's dealings with the Health and Safety Executive.

The company safety adviser must operate and be responsible for the annual budget for expenditure on health and safety activities. He must submit to the company safety director an annual report indicating

(a) the action taken and expenditure incurred on information, training, protective clothing and equipment, etc.

(b) incidents or accidents

(c) a general overview of the position of the company with respect to the fulfilment of its liability under the HSW Act.

The company safety adviser is required to have attended an approved course on construction safety for management and to maintain his knowledge of safety matters at least to the level required for membership of the Institution of Occupational Safety and Health.

The duties of the Resident Engineer. The term resident engineer is applied to the senior representative of the designing or specifying organisation on a construction contract. In most cases the resident engineer is the Engineer's Representative under the ICE Conditions of Contract but, in the context of health and safety, the responsibilities could apply equally to a resident architect or clerk of works.

As the senior representative of the employer on the site the resident engineer is responsible for the safety of staff and, in addition, is required to use skill and judgement in preventing any errors on or omissions from the drawings giving rise to hazardous situations. Ensurence that the employer's duties in law and the resident engineer's own are not compromised by the acts or omissions of the contractor is also expected(*see* Fig. 6.1).

A post entitled *engineer's safety adviser* is indicated in Fig. 5.4. This is an appointment by the resident engineer of one of the senior and experienced members of staff to take a special interest in safety matters at the working, as opposed to the managerial, level. This post has no particular authority under the contract or in law but it is found in practice that the nomination of a person who will, among other duties, co-operate with the contractor's site safety adviser and advise him of hazards of which he may not have been aware, brings about a good site safety discipline.

The summarised duties of a resident engineer with respect to health and safety are to

- ensure that working places and access to and egress from them are safe for himself and his staff
- ensure that he and his assistants have received suitable information and training in matters of safety; this includes knowledge of the HSW Act, the regulations relating to construction and other legislation and codes relevant to the works
- where appropriate, appoint an engineer's safety adviser, provide suitable terms of reference and inform the contractor
- assess the risks to staff on site, write them down and inform staff accordingly; provide personal protective equipment for staff and make sure they wear and use it properly
- ensure that his staff, and especially those with limited experience, are properly supervised
- ensure that first-aid and welfare facilities are available for his staff
- carefully follow all instructions on the drawings or in contract documents relating to procedures that affect the safety of construction workers or anyone else
- communicate regularly with the engineer's project manager and, in particular, discuss any details or instructions on the documents that require clarification
- administer the conditions of contract and especially the clauses relating to health, safety and welfare with due diligence; bring to the contractor's attention in writing any hazardous or potentially hazardous situation that could affect the health of anyone on or near the site; discuss site safety at regular progress meetings and record items for action.

Sometimes after the tender stage the originally appointed planning supervisor may be changed and the position given to the resident engineer or architect for the period of the construction phase. This will not affect his supervisory duties under the contract but in some circumstances for example when there are modifications to the design after site work has commenced, the appointment could involve duties that normally occur during the planning phase such as

- notifying the Health and Safety Executive of changes to the project
- modifying the health and safety plan
- giving advice to the client.

In taking over the duties of planning supervisor, the individual then becomes responsible for the health and safety file, adding to it as necessary and handing it over to the client on completion.

6 Management systems for safe construction

To satisfy the legal duty to provide for the safety of employees and others from risks arising out of work activities, it is necessary for every employer concerned with construction (client, professional adviser or contractor) to recognise the hazards and manage operations to eliminate them as far as is reasonably possible.

To achieve construction work that is safe to build, operate, maintain and demolish requires, in part, a responsible attitude by clients towards the selection of their professional advisers (consulting engineers, architects, quantity surveyors, etc.) and contractors, an acknowledgement of safety requirements in the preparation of design specifications, properly organised pre-contract investigations by engineers, the recognition of construction safety by designers and the systematic central management of safe methods by contractors.

There are two pieces of legislation that are aimed at ensuring that construction is effectively managed from a health and safety point of view. They are the

- Management of Health and Safety at Work Regulations 1992
- Construction (Design and Management) Regulations 1994 (*see* p. 20).

Management of Health and Safety at Work Regulations 1992

The MHSW Regulations apply throughout industry, including construction work, and must be complied with by employers, self-employed and employees. The main requirements of these regulations are for employers to

- assess the risks to their employees and anyone else who might be affected by the work, keep the assessment under review and modify as necessary. Employers with five or more employees must record the significant findings of their risk assessments; however even if there are less than five employees it is prudent to record the findings of a risk assessment in order to demonstrate to a Health and Safety

Executive inspector, for example, that a suitable and sufficient assessment has been made [similar duties are required of the self- employed]

- make and record the arrangements for the planning organisation, control monitoring and review of the preventative and protective measures identified as necessary by the risk assessment
- appoint one or more competent persons to help with implementing the health and safety arrangements
- provide for health surveillance where there is a risk of disease or other health damage
- set up emergency procedures for situations which might arise presenting serious or imminent danger
- provide information which can be easily understood about the risks involved and the necessary preventative and protective measures
- provide their employees with adequate health and safety training including refresher training
- work together with other employers in the same workplace to enable each other to understand all the risks associated with the project and to enable everyone concerned to comply with the relevant statutory provisions.

Details of employees duties, regulations concerning temporary workers and full details of the MHSW Regulations are contained in the approved code of practice, obtainable from HSE Publications (*see* Appendix 4).

Pre-contract activities

Initiative by clients

Safe workmanship is unlikely to stem from disinterested clients who are merely concerned with least cost and time and who give no encouragement to tenderers wishing to produce a quality job. There are good reasons why the client should take an active part in the promotion of safety on site, for example

- the publicity following serious accidents may tarnish the client's image
- the effects of resulting delays can seldom be fully recovered
- the client has a statutory duty to comply with the CDM Regulations (*see* p. 20 and Fig. 6.1)
- liability for damages suffered in the event of an accident.

Clients must appoint a 'planning supervisor' and later on a 'principal contractor' and be satisfied they are competent. They must provide the planning supervisor with all the information relevant to health and safety on the project and they must not allow work to start on site until the principal contractor has prepared a health and safety plan for the construction phase.

Fig. 6.1. Fatal accident due to a lack of clearance between roadworks and the live carriageway. The client and the contractors were found guilty under section 3 of the HSW Act (courtesy Derby Evening Telegraph)

Notification. If the project is notifiable then details of it must be provided to Health and Safety Executive by the planning supervisor as soon as practicable after his appointment and in any event before construction starts. A project is notifiable if construction will take longer than 30 days or involve more than 500 person-days on site.

The particulars to be notified to Health and Safety Executive are given in Schedule 1 of the approved code of practice to the CDM Regulations. Form 10 (rev) can be used to notify Health and Safety Executive of construction work, copies of which are available from the Executive.

Selection of contractors. Clients should set the standard for a safe project by selecting only those contractors, and those consulting engineers or other professional advisers, with acceptable safety policies to work or act for them and who otherwise meet the requirements of the CDM Regulations for competence and resourcing (*see* p. 20). This is done by calling for the policy statements and current arrangements for implementing them before signing the contracts and by making regular checks to ensure that those policies are effective. Clients should also check on the safety record of tenderers by requiring records of fatalities and reportable injuries for the recent three-year period and by rejecting from the list of tenderers those who fail to satisfy this scrutiny or fail to satisfy any requests of explanation.

The planning supervisor has a duty under the CDM Regulations to advise his client on the competence and resources of the designers and contractors. The reasonable steps that should be taken to assess competence are given in the Approved code of practice to the CDM Regulations but in general terms both designers and contractors need to demonstrate their knowledge, ability and resources to carry out the work and their policy and organisation to manage the health and safety and fire risk aspects of their work effectively.

Contract documents. The contract documents include among other things, the conditions of contract, the specification and the drawings. Instructions or information to the contractor may be shown on any of these. Whatever form of contract conditions are used, the client is advised to ensure they are compatible with statute law (*see* p. 31).

The pre-contract health and safety plan must be included among the tender documents and these may form part of the contract when it is signed. The health and safety plan must include as far as is known to, or forseeable by the planning supervisor the information a competent contractor might need to be able to take account of health and safety hazards. Examples of what should be included are given in Appendix 4 of the approved code of practice to the CDM Regulations.

In the contract documents, the names of the planning supervisor and where appropriate the principal contractor and their addresses should be stated. Compilers of contract conditions should consider the use of Clause 71 of ICE Conditions of Contract Fifth or Sixth Editions or an equivalent if some other form of contract is used (*see* p. 33). Instructions to the contractor from the principal contractor must be construed as coming from the client or his representative otherwise there will be managerial confusion which may lead to a loss of safety determination.

The authority of the resident engineer with regard to the construction phase health and safety plan should be addressed. Some clients will require the resident engineer to comment on and approve the health and safety plan for the purposes of the contract, thus retaining the mode of managerial control established by the ICE Conditions of Contract. Other clients may be content to leave all matters concerning the construction phase health and safety plan in the hands of the principal contractor.

The principal contractor has a number of tasks to perform during the progress of the work (*see* p. 25). These tasks are statutory duties under the CDM Regulations of which contractors are fully aware when submitting their tenders. It is unnecessary to repeat these statutory duties in the contract documents although the position and authority of the resident engineer in health and safety matters and his relationship with the principal contractor may require clarification.

Sometimes the potential for loss of life or production as a result of an accident is so great that the client needs to exercise a degree of direct authority in safety matters. In such cases, the powers and responsibilities of those concerned must be clearly defined in the specification or as special requirements to the conditions of contract. Other special requirements may be necessary where public utility companies etc. are affected by the works.

Special requirements

The public utility companies and other public and private undertakings that carry a special responsibility for public service and safety, issue special requirements for use in contracts where the apparatus or installations might be damaged. The most common of these are the public utilities whose pipes and cables are especially vulnerable, as are the workers who damage them! The special requirements are usually bound into the set of contract documents issued to all tenderers or otherwise made available to potential contractors during the negotiation stage. These special requirements must be included with the pre-tender health and safety plan. Compliance with the special requirements is a condition of contract and any default by the contractor is subject to the appropriate action by the engineer. The special requirements comprise a set of rules that

- describe the hazards
- tell contractors what, and what not, to do to avoid damage
- explain how to get the service cut off or temporarily suspended
- give rules for protection, marking, watching and supervision
- say what to do and who is the person to contact in the event of an emergency.

Standard special requirements for the British Isles are issued by the electricity generating and distribution companies, gas and oil distribution companies, waterways and rivers authorities, telecommunications companies and motorways communications systems and water and sewerage companies. Other special requirements are prepared or made available to compilers of contract documents whenever work is to be carried out on or near places of particular hazard such as airports, railways, coal workings and some industrial plants.

Survey and investigation

It is frequently necessary for surveyors and office-based engineers and architects to carry out topographic, geotechnical, hydrographic, structural or traffic surveys, to measure work or to study conditions under which the work will take place. Many of these site investigation assignments are hazardous and deaths and injuries have occurred to staff making them.

Every project must have its project manager or project engineer whose duties include seeing that staff on assignment out of the office have safe workplaces and means of access to them (*see* Fig. 6.2). The project

Fig. 6.2.   Articulated access platform in use for inspection and maintenance (courtesy Hewden Stuart Limited)

manager should make sure that those staff have received the appropriate instructions and training about the risks and hazards they are likely to encounter and in the ways of avoiding accidents. It should also ensure that they are provided with, and use, any protective clothing and equipment necessary.

The project manager's task is less onerous in companies that provide a good back-up safety organisation. Staff who have received basic safety training should be aware of the causes of accidents and how to avoid them. They should also be aware of the circumstances when protective clothing and safety equipment are required and where it may be obtained. The advantages of giving basic safety training to all staff as a matter of routine cannot be over-emphasised (*see* chapter 9).

Working alone

No individual should carry out an inspection, investigation or survey unaccompanied. In case there should be an accident another person should always be at hand to call for assistance. Engineers on site assignments should leave a message at their normal place of work indicating their whereabouts and expected time of return.

Companies should have rules on working alone as part of their safety policies and these would demand a written risk assessment as a pre-requisite (see below). The use of mobile telephones by lone workers is strongly recommended as is the use of personal alarms by women staff. Useful advice is available from the Suzy Lamplugh Trust, recognised by the Health and Safety Executive as giving the best guidance on lone working. The Trust issues a four point PLAN: 'Plan ahead, Look confident, Avoid risk, Never assume'.

Risks to surveyors and others on assignment

Management should consider the use of portable telephones by staff on survey, investigation or inspection assignments where other communications are unlikely to be available. This applies to geotechnical, hydrographic and traffic surveys and, for example, to the survey of unoccupied buildings.

Staff must always wear stout and sensible shoes or work boots as appropriate and, unless exempt by the Construction (Head Protection) Regulations, safety helmets. Other protective equipment may be needed depending on the assignment (*see* chapter 10).

Special care should be exercised when staff need to work at a height. It is often the case that office-based staff are not trained in fall prevention and safety equipment is not readily available for inspections which have to be carried out at height. However, staff should not be put at risk, nor should they themselves take risks, merely because the time or cost of working safely appears to be out of proportion to the perceived dangers, or their duration.

Other activities where office-based staff may be at special risk and where training and supervision are essential include

- surveys of buildings for demolition
- work adjacent to, or over water
- work on contaminated sites
- work on live highways and railways
- entry into places with excessive noise, dust or fumes
- sites where permits to work are required (*see* p. 218)
- work on roofs.

Staff who habitually carry out surveys and inspections of sites and structures are recommended to undergo basic first-aid training and to carry first-aid equipment with them (*see* p. 268). The cost and time involved in first-aid training is negligible compared with the benefits in the event of accidents, which, as we have seen, are not uncommon.

The MHSW Regulations require employers to formalise the method by which they assess and control the risks that their employees are required to face. It is therefore necessary for the project manager or other person in control of staff going on assignment to

- list the hazards at the site
- decide who is at risk
- list what measures are to be taken e.g.
  ○ to remove or to reduce the hazard
  ○ issue personal protective equipment
  ○ provide information, training, supervision
  ○ institute permits-to-work procedures
- review assessment and revise when necessary

Design and specification

The person responsible for the design of building or civil engineering works is normally one of a team of engineers, architects, graduates and technicians, each of whom is dependent on the other for producing safe, reliable and economic work, often to a predetermined budget. It is frequently assumed that, provided an engineer is qualified, the design produced will contain no mistakes and the work will be safe to build and operate.

In fact there are so many hazards, ranging from human error in calculations to lack of understanding of construction procedures and ignorance of science of the environment, that one engineer working alone is statistically unlikely to be able to carry out a design totally free from error. The preparation of designs, specifications and reports must, therefore, be properly managed so that the necessary level of expertise and experience is applied to all projects and so that mistakes are eliminated by an adequate system of checking.

Extent of legal and moral duties

The designer has both legal and moral responsibilities for the safety of the workforce and others. The moral responsibility of a chartered civil engineer is covered by the Institution of Civil Engineers' *Rules for professional conduct*: 'A member in his responsibility to his employer and to the profession shall have full regard to the public interest, particularly in matters of health and safety'.

The designer's criminal law duties are generally covered by section 3 of the HSW Act although there are circumstances where section 2 could apply. Where an article is being designed for use at work (e.g. a lifting device) section 6 could also apply. Specific regulations that cover the duties of designers are contained in the CDM Regulations (*see* p. 20). A designer would also be liable in a civil law action for damages suffered by reason of a faulty or incomplete design, specification or instruction. To what extent a designer would be liable for site accidents resulting from designs that are difficult to build safely would depend on the circumstances, but there is no doubt that any design or specification that fails to measure up to the requirements of the CDM Regulations may invite

Fig. 6.3. Facade retention is a technique requiring close co-operation between designer and contractor (an example in Brussels)

criminal proceedings and of course will be difficult to defend in any civil action alleging negligence.

Complying with the CDM Regulations

There are a number of specific duties required of designers in complying with the CDM Regulations, for example

- providing evidence of competence and resources
- ensuring the client is aware of his own duties
- passing information to the planning supervisor
- co-operating with the planning supervisor and any other designers
- considering hazards and risks, designing to avoid the risks and protecting workers where the risks cannot be satisfactorily reduced.

Taking account of the hazards of construction and minimising the risks to health or safety of those who are going to construct, maintain, clean, repair or demolish the structure requires a great deal of knowledge of safe construction methods and of the causes of accidents and occupational ill health. A comprehensive understanding of the contents of chapter 4 of this book is essential if the designer is to cover his duties under the CDM Regulations effectively.

Designers should be aware of the heirarchy of risk control which under-lies the modern approach to health and safety management. It is best to avoid risks by effective design. If this is not reasonably practicable, the risks should be combated at source. Failing this, priority should be given to controls that will protect all workers and only then should personal pro-tective equipment be assumed. Guidance for designers involved in con-struction work is given in *Designing for health and safety in construction*, a guide for designers on the CDM Regulations published by the Health and Safety Executive (*see* Appendix 4).

Design aids for safe construction

The number of ways in which a designer can contribute to safe construction are too numerous to cover here. Nevertheless, an understanding of the causes of site accidents by designers will help to focus their attention on construction details which are potentially dangerous. Since a substantial number of fatalities and injuries result from falls it is important that designers should reduce the need for people to work at a height to the absolute minimum. This applies to construction workers and also to opera-tional and maintenance staff. By reducing the number of people working at a height, for example, by maximising pre-assembly of frames at ground level (Fig. 4.17), a contribution is also made to reducing the number of falling pieces of equipment and materials, which is another major cause of injury and death.

It is likely that designers will sometimes face a dilemma where the safest method of construction is not the cheapest. Their responsibilities to the client and to the contractor's workforce will not always be in accord. Items of an exceptional nature where the cheapest method of construction carries risks must be discussed with the planning supervisor and the client. This is another area in which clients can contribute to site safety. A simple example is shown in Fig. 4.32.

The Construction Industry Advisory Committee's *Guide to managing health and safety in construction* published by the Health and Safety Execu-tive (*see* Appendix 4) lists a number of items commonly encountered on construction sites, information on which should be included in the pre-tender health and safety plan. They are

- surrounding land uses and related restrictions, e.g. schools, shops or

factories adjacent to the proposed site, planning restrictions which might affect health or safety
- existing services, underground or overhead lines
- existing traffic systems and restrictions, e.g. access for fire appliances, ease of delivery and parking
- existing structures, e.g. special health problems from materials in existing structures which are being demolished or refurbished, any fragile materials which require special safety precautions, or instability problems
- ground conditions, e.g. contamination, instability, possible subsidence, old mine workings, underground obstructions.

The limits of available working space could be added to the list (*see* Fig. 6.1).

The pre-tender health and safety plan should include drawings of any structures to be demolished. It should also include drawings, diagrams or clear instructions about the sequence of assembly of any structures where there is a risk of instability or unwanted built-in stresses (Fig. 6.3). Guidance on the contents of the health and safety plan is given in the Approved code of practice to the CDM Regulations (*see* Appendix 4).

During the progress of the works, the value of good communications between the design team and the resident engineer or site manager cannot be over-emphasised. Instructions on the drawings or in specifications, or sometimes the lack of instructions, can have serious consequences where they are unclear and where resulting queries are unsatisfactorily resolved. A system of communications should be set up before construction starts (*see* Fig. 6.4). The need for a comprehensive and clear health and safety plan is obvious on complicated high-risk projects, as illustrated in Fig. 6.5.

Design aids for safe maintenance

It is important for the designer of most civil engineering and building works to be thoroughly aware of the need for, and methods of their maintenance. This too is a duty on designers under the CDM Regulations. In some cases, for example in sewer works, the requirements for maintenance are laid down by the adopting authority, giving the designer guidance on the provisions that must be made in the layout and detail design of manholes. Reference should be made to *Sewers for adoption: a design and construction guide for developers.*[1] However, for a great many large building and civil engineering works, provision for their safe maintenance must be the subject of discussions between the designer and the eventual owner or maintaining authority, preferably before design begins.

Maintenance is a hazardous occupation. The Health and Safety Executive estimate that an average of 43% of all construction industry fatalities in the five years 1981–85 occurred during maintenance, two-thirds of which were due to roofwork and other falls.

Fig. 6.4. Damage to an underground train caused by soil sampling apparatus for highway works (left) that had penetrated the tunnel. Poor communication had failed to resolve an anomaly on the borehole setting-out drawing.

Promoters of construction works should provide funds for, and designers should specify safe means of access to places at a height where maintenance staff are required to work. For buildings, the Building Regulations 1985 (*see* Appendix 2) provide for the safety of those who occupy or use buildings. Stairways, ramps and guards are required to be erected to given minimum standards at places where people normally have access. However, for the majority of civil engineering and building works the provision of safe access for occasional maintenance should comply with the CHSW Regulations, or relevant Health and Safety Executive guidance notes (*see* Appendix 4).

Maintenance of civil engineering works is as diverse a subject as civil engineering itself but a few examples will show how a designer and client, owner or operator can assist in the reduction of maintenance hazards.

Choice of materials. If building material is selected with a view to future maintenance, then the actual need for maintenance can be reduced, in some cases to negligible proportions. The repainting of a steel highway bridge, for example, is not only hazardous in itself but also lane closures impair traffic safety.

Fig. 6.5. Careful site planning is necessary to ensure compliance with the safety plan on contracts where departure from the designed sequence could be catastrophic (construction of Limehouse Link Tunnel, London Docklands – courtesy Balfour Beatty Amec Joint Venture)

The need for maintenance, in this example, can be reduced by the use of reinforced or prestressed concrete. The soffits and sides of concrete bridge spans need (or should need) negligible maintenance. However, a maintenance-free structure is not always the cheapest and the designer should present the options to the client.

Shape of structures. Where ledges, cavities, joints, internal angles and other corrosion-inducing features and dirt traps are eliminated, the need for cleaning and repainting is reduced. Box members require less maintenance than plate girders and lattice frames and, in addition, the hazards to maintenance workers are more easily eliminated.

There are many other similar instances where the designer can reduce the need for maintenance and the hazards of carrying it out by careful attention to the exposed outline of the structure and to the detailing of its parts. Expansion joints, for example, give considerable maintenance problems, especially in bridges, and should be avoided at places of difficult access. Again, the designer's proposals may not always be the cheapest initial solution and discussions with the client are necessary.

Corrosion protection. The choice of corrosion protection system influences the maintenance interval. Generally the selection of corrosion protection is based entirely on economics but the designer should ensure that the cost of safety provision for maintenance workers is taken into account.

Provision for safe maintenance. Maintenance hazards can be considerably reduced in major structures by providing safe access by means of permanent purpose-made power-operated cradles, passenger and goods hoists, platforms and gantries. These are commonly provided for major bridges, large roofs, chimneys, silos and large buildings. Where permanent maintenance platforms are impracticable or where they fail to provide complete protection the designer must consider the safe working of maintenance crews by specifying ladders, step irons, handholds, anchorage points for use with safety harnesses and fixing eyes. Ladders, tower scaffolds, man-riding hoists, etc., and their employment, must also be considered.

The Joint Advisory Committee on Safety and Health in the Construction Industry drew attention in 1973 to the need for designers of buildings to provide means for erecting and tying scaffolding for future maintenance purposes. Even for existing structures not initially provided with tying positions for scaffolding, consideration should be given to permanent anchors drilled in or otherwise secured during the first maintenance programme.

The safety of roof workers deserves special mention since they suffer the highest injury rate of all maintenance staff. Wherever possible, access to roof areas should be by permanent, fixed ladders fitted with hoop guards but, where permanent ladders are impracticable, anchorage points for portable ladders should be provided to enable them to be securely tied (Fig. 4.21). Permanent walkways with guard rails and toe boards should always be provided wherever regular maintenance is necessary at roof level. For work on fragile roofs there are advantages if the permanent walkways are installed above the fragile covering to enable maintenance to be carried out to roof lights, gutters and ventilators. On multi-bay structures, it is sensible to provide access over the ridges. Invariably, access will also be required beyond the walkways and, wherever this is the case, permanent attachments for safety lines should be provided.

Safe access should also be designed for the maintenance of the structure and items in the roof space, especially above suspended ceilings in industrial and commercial buildings. Where permanent ladders and walkways are not viable, provision for securing temporary ladders and for anchoring temporary walkways and staging should be made. The provision of suitable eyes to which safety lines can be attached should also be considered so that when work is undertaken at heights, workers can operate in a safe manner (Fig. 6.6). The hazards of roofwork and their prevention are covered in greater detail in chapter 4.

Design aids for safe demolition

For the planner of demolition activities, the availability of accurate as-constructed drawings is vital. On completion of every project, the planning supervisor must issue to the client a health and safety file containing copies of the construction drawings modified by site revisions, giving exact details

Fig. 6.6. Travelling staging making use of crane rails to provide a working platform for the underside of a large-span roof

of the works as constructed. These must be retained by the client and made available for anyone who might need the information. The original as-constructed drawings should also be retained indefinitely by the design organisation, preferably on microfilm in safe storage, and properly ar-chived. When modifications to existing structures are required, when total demolition is contemplated or where there is evidence of failure or distress, the availability of accurate as-constructed drawings and calculations will reduce the risk to people engaged in surveys and demolition tasks.

There are some permanent structures that the law requires to be re-moved. They include offshore oil and gas installations. Other permanent structures (e.g. nuclear reactors) will have to be demolished within a foreseeable time span and their design should take into account not only the physical dangers to demolition workers but also the probable health hazards from contaminated materials.

Most building and civil engineering works, however, are designed without regard to their eventual demolition. Nevertheless, every project has its special circumstances and the designer is advised to discuss any possibility of partial or total demolition with the client or promoter. A study of British Standard BS 6187 *Code of practice for demolition*[3] and HSE guidance notes GS 29/1, /2, /3 and /4 (*see* Appendix 4) will give the designer advice on planning, preparation, legislation and techniques for safer demolition. Advice is given on the hazards of some special demolition work which the designer may be able to alleviate in the original design. It includes prestressed concrete chimneys, kilns, arch viaducts, jack arch floors and steel storage tanks.

Safe systems of work for design offices

The design of all building and civil engineering works should be carried out by or under the close supervision of professional staff thoroughly competent in the relevant technology. All designs must be checked and the calculations and drawings signed by the checker as well as the designer and technician. The level of checking must be decided by the responsible senior engineer, depending on the importance of the work and on the degree of risk.

Arithmetical checking. An arithmetical check of all calculations must be made and every drawing must be checked to see that the intentions of the designer have been correctly interpreted and indicated. While the arithmetical and visual check may be carried out by an engineering graduate, the fundamental check must be made by an experienced engineer.

Fundamental checking. A fundamental check of the design and drawings is necessary to ensure that the correct design methods have been employed, that the relevant codes of practice have been used and that the safety of those who build, operate, maintain and demolish have been taken into account. The checker must be sure that the work can be constructed safely and that any instructions indicated on the drawing, or in the specification, are both correct and unambiguous for the contractor. A secondary check of critical calculations should be made by a different method from the one employed in the original design as a further verification of the design philosophy.

Design reviews. Depending upon the extent and complexity of the project it is advisable, and in many cases necessary, to review the design process from time to time. The purpose of design review is to identify any hazards brought about by the design and to effect any necessary changes. Additional purposes are to review resources and to co-ordinate the work of designers on the project. The review should be conducted by the design team leader, the project manager or the planning supervisor as appropriate, and minutes should be taken and circulated for approval.

Independent checking. Where civil engineering or building works whose failure in construction or use might cause multiple loss of life it is advisable (indeed it is required by authorities responsible for certain structures such as highway bridges, dams and offshore structures) for a totally independent check of the design to be made. Some authorities will allow this check to be carried out by the same organisation as that which did the original design but by a different design team. In other cases the independent check

---

CONSULTING ENGINEERS
RAILWAY PROJECT QUALITY PLAN

SECTION X - DESIGN CONTROL

CLASS 1 - DESIGN CERTIFICATE

1. We certify that reasonable professional skill and care has been
    used in the preparation of the design of:

    with a view to securing that:-

    i) It has been designed in accordance with Railway
       Project Design Criteria.

    ii) It has been checked for compliance with the relevant
        standards in i).

    iii) It has been accurately translated into Drawings which
         have also been checked. The unique numbers of these
         Drawings are:-

    Signed..(Team Leader)

    Name ....

    Signed..(Director)

    Name ....

    Date ....

2. The Certificate is accepted by The Railway Project Authority

    Signed...

    Name ....

    Date ....

Fig. 6.7.   Example of design certificate (courtesy W.S. Atkins Consultants)

must be performed by a different company. A typical independent check
certificate is indicated in Fig. 6.7.

The importance of continuing education. The importance of persistent
reading about the causes of accidents and failures emanating from the
design office cannot be over-emphasised. This can be assisted by the
circulation of journals and other literature from a library or other central
source. The wisdom and experience of senior engineers retiring from the
profession has to be replaced by the continuing education of younger
engineers to pass on the lessons learned from the hundreds of construction
site accidents and other tragedies that are reported every year. Attendance

at seminars and conferences organised by the Institution of Civil Engineers and others on all health and safety matters must be encouraged. Designers are unable to demonstrate their competence to act under the CDM Regulations if they have insufficient knowledge of the causes of site and maintenance accidents and ill health and the measures available for overcoming them.

Quality assurance in design

An effective design office system to produce safe and reliable designs and specifications is the principle of quality assurance. This principle is the subject of CIRIA report 109 *Quality assurance in civil engineering*, published in 1985.[4] Quality assurance implies the application of management techniques to

- assure reliability of public safety by structured control of all stages of design and construction
- provide positive evidence that the quality produced by the design and construction meets that specified
- demonstrate prudent expenditure.

In civil engineering design, therefore, quality assurance goes beyond that required for reliability or safety but its adoption will certainly ensure that the works have been designed and constructed to principles that will result in their being reliable and safe.

Even without the application of quality assurance, the establishment of a project organisation with its management statement, controls and audits will ensure, as far as is practicable, that the project is error free. A design office in which a quality assurance policy has been adopted requires a quality management manual. Every project conducted to quality assurance principles requires a project quality plan.

Quality management manual. The quality management manual is a statement of policy by the design office and defines the intent, organisation and responsibilities of the quality management functions of senior staff. The senior staff must be named individually and their reporting lines clearly stated. The quality management manual sets out the procedures covering all quality-related aspects of the design office work. It describes in detail the actions required to create a project quality plan and covers monitoring, records and staff training. All phases of design from preliminary studies to design and calculations to inspection of construction are covered in the quality management manual. Names of the company quality assurance director, the quality managers and the company safety adviser are given, together with names of the chief engineers responsible for the various technologies. Fig. 6.8 shows a company structure of responsibility for quality assurance for a firm of consulting engineers.

The reporting line of the planning supervisor under the CDM Regulations is also shown. The planning supervisor may not be a company employee, he may be appointed separately by the client. If the company is

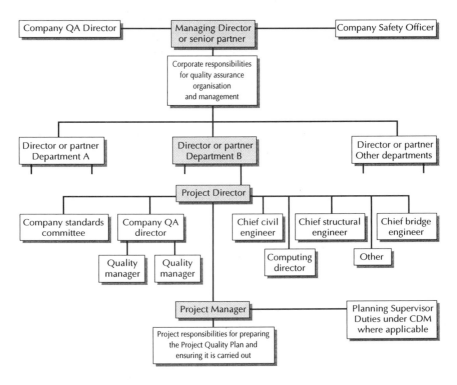

Fig. 6.8. Consulting engineers' structure of responsibility for quality assurance (courtesy W.S. Atkins Consultants)

appointed as planning supervisor then it is essential to allocate the tasks to the project manager or to another individual.

Project quality plans. Every project to which quality assurance principles apply must have a unique project quality plan. Among other things it describes the organisation, responsibilities of staff on the project, document control, design control, quality audit and review and site supervision. These are each described in turn.

The first section of a project quality plan, the *organisation*, describes the purpose of the project, the companies involved and their contractual relationships to one another. It describes the management structure and identifies all the senior staff with project and technological responsibilities. Figure 6.9 illustrates the quality assurance organisation for the design of the civil engineering works for the London Docklands Light Railway Project.

The next section provides a clear brief for every group or organisation involved in the project and describes the *responsibilities of senior staff* relating to the project. The section has to be compiled carefully so that no tasks are left undone and so that no member of the design or checking team is in any doubt of his or her reporting line.

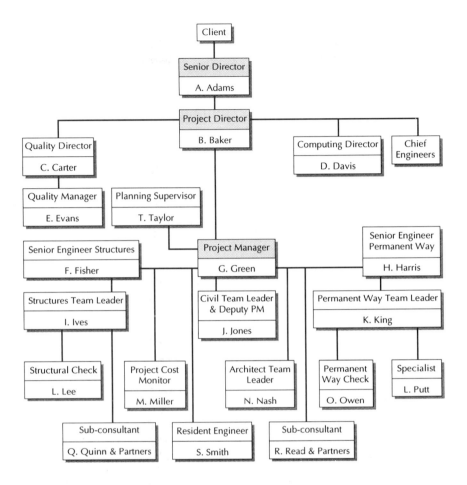

Fig. 6.9. Quality assurance organisation for the design of the civil engineering works for a railway project (courtesy GEC–Mowlem Railway Group)

For the proper management of a large and diverse project it is necessary to publish the *procedures for communication*. Documents include mail, telex, minutes of meetings, reports and drawings. Distribution networks and standard file references must be devised to reduce the risk of errors due to misdirected information. Figure 6.10 gives an example of the routine for dealing with incoming mail.

Design control. Design control is concerned with the engineering procedures for the technical departments engaged in the project. It lays down the

- design standards
- specification and check procedures
- methods of producing design calculations and performing design certification

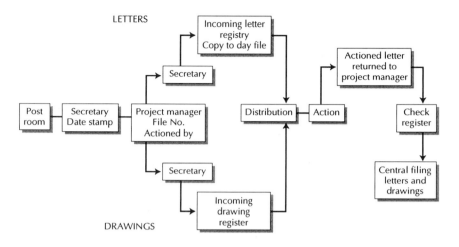

Fig. 6.10. Procedure for dealing with incoming mail

- method of producing drawings and check certification
- control of changes to the design and drawings.

A typical design certificate is illustrated in Fig. 6.7.

To ensure that the requirements of the project quality plan have been followed and that the documentation and record systems are adequate, it is necessary to carry out a *quality audit* at regular intervals. The audit is carried out by a quality manager in the full-time presence of the project manager to an agenda that follows the audit checklist (*see* Fig. 6.11). The audit report is issued to the project director, the quality director and the project manager, who takes any necessary corrective action. It is also issued to the planning supervisor for inclusion in the safety file.

Although not strictly a design office activity, the project quality plan must describe the responsibilities of the *resident engineer*, if there is to be one, and clarify his reporting line with respect to the project manager, the principal contractor and others responsible for the technical aspects of the work. Resident engineers should receive instructions on procedural matters, on health and safety aspects of the work and on their responsibility for the health and safety of staff appointed to assist them. If the construction is also to be quality assured there may be no resident engineer (see later in this chapter).

Management of construction

Safe systems of construction are achieved by the application of effective management techniques together with the facilities and motivation to fully implement the published safety policy (described in chapter 5). So that

CONSULTING ENGINEERS RAILWAY PROJECT QUALITY PLAN SECTION Y – QUALITY AUDIT AND REVIEW				
A B C & PARTNERS London, England			Tel:	
CLIENT				
PROJECT				
QUALITY ASSURANCE CHECK-LIST		date: / / 9		
DISCIPLINE AUDIT		Report Ref No. AR		
DISCIPLINE: _____				
AUDITOR(S): _____				
PLANNING SUPERVISOR: _____				

Ref. No	CHECK ITEM	YES	NO	REMARKS
1.0	STAFF			
1.1	Have all staff signed as having studied required documents			
1.2	Are all staff approved and are CVs available in the Register of staff?			
1.3	Are staff familiar with required section of Project Quality Plan?			
1.4	Are staff familiar with the Procedures Manual?			
1.5	Are staff familiar with other relevant documents, e.g. Design, Criteria and Philosophies?			
2.0	DOCUMENTS			
2.1	Have documents issued been produced in accordance with Procedures?			
2.2	Are documents correctly filed and used, including those in other areas?			
2.3	Are documents in course of production in accordance with Procedures?			
2.4	What is programme of further document production? Is there enough time being allocated for QA?			
CHECK-LIST COMPLETED BY:		SHEET NUMBERS OF SHEETS		

Fig. 6.11.   Example of an audit checklist

management can be sure that its construction sites are properly organised, it is useful to have a standard system for setting up a site safety organisation and for monitoring its effectiveness. This may be effected by completion of a safety appraisal for every site, followed by safety audits from time to time during construction.

Safety appraisal

The site manager is responsible for setting up the safety appraisal when the contract has been awarded. The safety appraisal is the formalised statement of responsibilities and actions required for the safe and proper execution of the contract. It includes the names of staff with managerial and functional duties and a list of the primary operations to be carried out requiring action by the staff nominated. The responsibilities of subcontractors and their senior staff must be included.

When the safety appraisal has been prepared, meetings are attended by the site manager, safety adviser, line managers and relevant subcontractors to discuss the actions, identify the hazards and allocate responsibilities in order to devise methods to eliminate or minimise the hazards in relation to the construction programme. A typical safety appraisal is shown in Fig. 6.12.

Construction phase safety plan

Information derived from the safety appraisal is to be provided for the principal contractor to develop the safety plan required by the CDM Regulations (*see* p. 20).

The principal contractor has to state the

- arrangements for ensuring the health and safety of all who may be affected by the construction work
- arrangements for the management of the work and for monitoring compliance with health and safety law
- information about welfare arrangements.

It is therefore necessary for the principal contractor to be provided with

- details of resources
- evidence of competence to carry out the work
- risk assessments as required by the MHSW Regulations
- evidence of training
- evidence of information
- evidence of consultation with workers
- evidence of site security
- copies of RIDDOR reports.

Safety audits

It will be evident that the safety appraisal is meaningless if actions required are not carried out. On sites there is pressure to ignore safety procedures to save time or to improve productivity bonus payments. It is necessary for safety audits to be carried out at intervals by the company safety adviser or supervisor. The intervals between visits to a site by the company safety adviser will depend on the degree of hazard and the experience of the site manager. The degree of hazard will in turn be influenced by the seriousness of any possible accident, especially where the public might be affected, and by the pressure of a tight construction programme. Some site managers can be relied on more than others to organise a safe site. The safety audit is normally typed in a 'house style'. Pro formas have been found to be unsatisfactory.

Constructwell Limited
Safety Appraisal

Contract   Concert Hall

Appointments   Appointments contd.

Site manager	A. Adams	Compound	H. Harris
Temporary works co-ordinator	B. Baker	Registers:	
Temporary works inspector	C. Carter	Scaffold	C. Carter
Principal contractor	D. Davis	Excavators	C. Carter
Site safety supervisor	D. Davis	Cranes	I. Ives
Safety representatives		Hoists	I. Ives
First-aid attendants	E. Evans	Fork lift trucks	C. Carter
Accident book	E. Evans	Young persons	J. Jones
Fire officer	F. Fisher	RIDDOR	C. Carter, E. Evans
Welfare facilities	G. Green	Record of trained persons	J. Jones
Form 43B	E. Evans	Permission to break live sewers	K. King

Signed ..A. Adams..................................
Site manager

Date...

This appraisal takes effect from ....

Programme Item No.	Operation	Action req'd	By	Competent person (for register)	Comments
3	Erect tower crane	Plant dept. to provide erection details	Site manager	TC driver	Inspections F. 91 C - E
		Test certificate	Plant dept	Plant dept	Inspections F.91 G
		Drivers certificate	Safety dept		PN 52 D
		Banksmen & slingers training	"		Training req'd.
		Test certs. for S.L.I. & lifting tackle	Site manager	TC driver	Inspection F.91 J
7	Excavate pile caps & founds	All excav. to be battered back or timbered	Site manager	Foreman of excav	Inspection F. 91 Part 1B
8	Cut off pile tops	All ops to be issued with goggles	Site manager		Protection Eyes Regs.
11	Drainage	All excav. to be guarded, timbered & have safe access	Site manager	Foreman of excav.	Company Practice Notes 113D and 148 D Inspections F.91 Part 1B

Fig. 6.12. Example of a safety appraisal (courtesy John Laing Construction Ltd)

The audit report is circulated to relevant line managers but the responsibility for any corrective action lies with the site manager. It may be necessary for the safety adviser or even the construction manager to follow up a safety audit to ensure that any unsafe situations or practices have been eliminated. The audit report should also be copied to the principal contractor for CDM Regulations purposes.

Permits to work

Several operations are potentially so hazardous that normal training and supervision are insufficient to ensure safety. The permit to work is a document which describes the work to be done and the precautions to be taken so that the errors, omissions and misunderstandings that cause accidents are minimised, even eliminated. It provides a record that all foreseeable hazards have been considered by responsible personnel.

Operations for which permits to work should be used include

- entry into confined spaces, closed vessels and vessels containing agitators or other moving parts
- work involving the breaking of pipelines or opening of plant containing steam, ammonia, chlorine, other hazardous chemicals and hot substances or vapours, gases, or liquids under pressure
- work on certain electrical systems
- welding and cutting operations in areas other than fabrication shops
- work in isolated locations, locations with difficult access or at a high level
- work in the vicinity of, or requiring the use of, highly flammable explosive or toxic substances
- work which may cause atmospheric pollution
- work involving ionising radiation.

The use of permits to work is influenced by the degree of risk to which engineers, workers, members of the public, property and product are exposed in respect of

- the type of work undertaken
- the working method used
- the location of the work
- any articles and substances used which may affect, or be affected by, the work.

The permit-to-work system requires rigorous adherence to the arrangements and in particular the authority to specified persons to issue permits to work, the signing of the permits by authorised staff and the distribution of copies of the signed permits. An example of a permit to work is given in Fig. 6.13.

PERMIT TO WORK
ON
ELECTRICAL APPARATUS

Serial No. _____
Key Safe No. _____

1. **ISSUE**

To _____ , being a Competent Person.

In the employ of _____

I hereby declare that it is safe to work on the following H.V. Apparatus which is dead, isolated from all live conductors and all H.V. conductors are efficiently connected to earth.

_____

ALL OTHER APPARATUS IS DANGEROUS

Points at which the system is isolated and caution notices are placed:

_____
_____

Additional caution notices placed at:

_____
_____

The apparatus is efficiently connected to earth at the following points.

_____
_____

Dangers Notices/Barriers placed at:

_____
_____

Special precautions to be taken by the Competent Person in charge of the work:

_____

Other relevant permits issued:

_____

Key Safe No. _____

For the following work to be carried out:

_____

No. of additional permits issued: _____
Signed _____ Time _____ hrs. Date _____
being a Senior Authorised Person (E) authorised to issue a Permit to Work.

2. **RECEIPT**

Note: After signature for the work to proceed this receipt must be signed by and the Permit to Work retained by the Competent Person in charge of the work until the work is suspended or completed and the Clearance section 3 has been signed.

I hereby declare that I accept the responsibility for carrying out the work on the apparatus detailed on this Permit to Work and that no attempt will be made by me, or by the persons under my control, to carry out work on any other apparatus.

Signed _____ Time _____ hrs.Date _____
being the Competent Person to whom this Permit to Work is issued.

3. **CLEARANCE**

I hereby declare that the work for which this Permit to Work was issued is* suspended/ completed and all persons under my charge have been withdrawn and warned that it is no longer safe to work on the apparatus specified on this Permit to Work and that gear, tools and additional earthing connections are now clear.

Signed _____ Time _____ hrs. Date _____
being the Competent Person to whom this Permit to Work is issued.

4. **CANCELLATION**

I hereby declare that I have checked _____ Additional earth connections and that this Permit to Work is cancelled.

Signed _____ Time _____ hrs. Date _____
being a Senior Authorised Person (E) authorised to cancel a Permit to Work.

5. Additional information or variations*

_____

*Complete as required.

Fig. 6.13. Example of a permit to work (courtesy Thames Water plc)

Competent person

Tasks such as, for example, undertaking hazardous inspections, issuing permits to work or requiring exposure to a high level of foreseeable risk at frequent intervals should be undertaken only by specifically trained engineers or operators who appreciate the risks involved.

The expression 'competent person' occurs frequently in construction safety legislation. For example, under the CHSW Regulations inspections, examinations, operations and supervisory duties must only be undertaken by competent people. The onus is on the employer to decide whether people are competent to carry out these duties. An employer might do this by reference to the individual's training qualifications and experience. Broadly speaking, a competent person should have practical and theoretical knowledge as well as sufficient experience of the particular machinery, plant or procedure involved to enable identification of defects or weaknesses during plant and machinery examinations, and to assess their importance in relation to the strength and function of that plant and machinery. A competent person must be able to discover defects and to determine the consequences of such defects.

A competent person does not have to be so competent that he never makes a mistake. In a decision in the High Court in 1954, where the competence of a quarry foreman was in question following a rock fall from the face of a quarry, the judge held that the foreman was competent despite the apparent lapse in his inspection. The judge said 'I am not prepared to hold that competent means the most competent person available . . . or that it means he shall be so competent he never makes a mistake. A competent person is one who, on a fair assessment of the requirements of the task, the factors involved, the problems to be studied and the degree of risk of danger implicit, can fairly, as well as reasonably, be regarded as competent to perform such inspection'. He also added on the subject of competence that 'experience is so often of much greater value than book learning'.

Competent persons are involved in many activities, including

- supervision of erection or any substantial addition or alteration (dismantling) of a scaffold
- supervision of personal suspension equipment
- inspection of scaffolds before use, every 7 days, after alteration, etc, and after any event that might affect stability
- supervision of temporary works
- supervision of demolition or dismantling
- supervision of cofferdams and caissons
- supervision of work requiring training, technical knowledge or experience
- inspection of excavations, cofferdams and caissons at the start of each shift and after any event that might affect stability.

The word 'competent' is also widely used in the CDM Regulations in connection with individuals or firms appointed as planning supervisors, principal contractors, designers or contractors.

Quality assurance in construction

CIRIA report 109, *Quality assurance in civil engineering*,[4] states that one of the reasons that quality assurance is necessary is to assure reliability of public safety by structured control of all stages of design and production. It is evident that a contract that is quality assured will satisfy the requirement of the HSW Act that safe systems have been adopted.

It is not possible to draw any correlations, let alone conclusions, from the construction sites using quality assurance compared with those not doing so. However, it is axiomatic that safety thrives in the company of efficiency, high morale, tidiness, discipline and skill. These factors add up to quality.

It is common practice in large construction companies to monitor the accident records from various sites and to regard them as a barometer of the overall health of the job. The operation of quality assurance on construction sites has more often been applied to design-and-build contracts for the petrochemical and nuclear industries than for normal civil engineering contracts.

A typical quality plan for a civil engineering contract where quality assurance has been adopted is given in the following paragraphs. The contract was not in accordance with the ICE Conditions of Contract and there was no engineer. The supervision was covered by the quality assurance plan and a separate organisation was appointed to monitor the quality plan and act as principal contractor to satisfy the requirements of the CDM Regulations.

Construction quality plan. Quality assurance applied to construction is a system that effectively and economically ensures and demonstrates that workmanship and materials conform to the specified requirements. The quality assurance programme is aimed at providing a disciplined approach to all site activities affecting quality, including verification that each task has been satisfactorily performed and that the required level of quality has been achieved.

A quality assurance programme differs from the traditional construction system, which generally relies on the experience and integrity for foremen and inspectors supported by their engineers. A quality assurance plan does not require an engineer or an engineer's representative or his site staff. The construction manager, whose duties are outlined below, is responsible for the safe and efficient construction of the works, and an independent checker is engaged to see that the quality assurance programme is being carried out. The duties of the construction manager include

- ensuring the safe and efficient construction of the works
- ensuring that the quality assurance programme is complied with.

Quality control of workmanship. To ensure that the civil engineering works are built to the specification and drawings, quality checks are carried out at strategic stages in the construction. A strategic stage is that where a significant change of circumstances occurs such as placing concrete in the completed formwork. The quality checks are carried out only by people who have been authorised in writing and each check is properly recorded before the next stage of work can proceed.

Flow sheets. Each stage of the work from start to finish of the job is identified and briefly described on flow sheets. They stipulate those strategic stages at which quality checks are to be carried out and who is authorised to carry out the checks. An example of a flow sheet for a railway bridge excluding piling is given in Fig. 6.14. The quality control check sheets referred to are collected and distributed to the client and the independent checker.

Quality assurance review meetings. Before the start of any new form of construction or any major section of work, e.g. a bridge or a tunnel, a quality assurance review meeting is held. The meeting will be attended by the construction manager and the senior quality assurance engineer of the independent checker acting as principal contractor together with engineers and foremen responsible for the particular section of work. In some circumstances the planning supervisor and or the senior designer should attend the review meeting to ensure that any construction methods required by the design are fully understood and complied with.

The agenda of the quality assurance review meeting will be

- description of the work
- construction method and sequence
- review of flow chart and quality control checks to be made
- personnel authorised to carry out the quality control checks
- design check certificates required
- any other business.

The minutes of the meetings will be signed by those present and circulated according to the pre-arranged list. They will form part of the quality assurance records.

Non-compliance with the contract or quality control procedure. In the normal course of events any non-compliance with the drawings or specification will be identified by the engineer responsible for carrying out the quality control checks. The engineer will give instruction for rectification, and must then ensure that the work is properly carried out; there must be no signing of the control form until everything is in accordance with the contract.

When the independent checkers find that work is not in accordance with

Constructwell Group Railway Project		Date:...........................	
Quality Assurance Plan Section 5 Workmanship		Revision:.....................	
		Page:..........................	
Flow sheet Viaducts and bridges		Fig	

No.	Activity	Check	Grade of staff to check	Control Form No.
01	Excavate pile cap and clean ready for blinding	QC check	11	QA5
02	Place blinding, prepare pile tops, survey positions	Survey	13	
03	Fix shutters, reinforcement kickers	QC check	11	QA6
04	Concrete pile cap and cure	Concrete check	13	QA17
05	Fix column shutters and reinforcement	QC check	14	QA7
06	Concrete column and cure	Concrete check	13	QA17
07	Prepare column top for bearing and check position	QC check	13	QA11
08	Fix bearings and grout	QC check	14	QA11
09	Place permanent steelwork, bolt and weld	QC check	13	QA12
10	Fix deck shutters and reinforcement	QC check	14	QA20
11	Concrete deck and cure	QC check	13	QA17
12	Prepare deck for waterproofing	QC check	13	QA21
etc	etc	etc	etc	etc

Fig. 6.14. Example of a quality assurance flow sheet (courtesy GEC–Mowlem Railway Group)

the drawings or specification or that the agreed quality control procedures have not been followed, they issue a non-compliance certificate to the contractor, who must sign that it has been received. A typical non-compliance certificate is given in Fig. 6.15.

On receipt of a non-compliance certificate, the contractor must take immediate steps to identify the nature of the non-compliance. The contractor will then

- rectify the work, or
- define the method and timing of the remedial work, or
- ascertain from the designers whether the non-compliance is acceptable and, if not, rectify the work.

When the non-compliance has been rectified, the non-compliance order is revoked by the independent checker, but if the order has been ignored, the independent checker must report the matter to the client. The client, of course, will have safeguards under the contract to enable him to apply

Quality Inspections Group	Revn.	
Railway Project	Date	
	Page	

Quality Assurance
Non-compliance notice No.....................................

1

The workmanship/materials for the section of works described in Section 2 below do not comply with the contract for the reasons given in Section 3 below. Please rectify the matter forthwith.

Signed ....   Received..

Quality Inspections Group   Constructwell Limited

Date.................. Time..................   Date.................. Time................

2

Section of Works
Reference number
Particular part of work not complying

3

Details of non-compliance

4

Rectification measures
[To be completed by contractor]

5

Because the measures set out in Section 4 above have been taken it is agreed that the faults set out in Section 3 above have been rectified

Signed ....
Quality Inspections Group

Date..

Fig. 6.15. Example of a quality assurance non-compliance certificate (courtesy W.S. Atkins Consultants)

financial pressure on the contractor if he fails to comply with the quality procedures.

Any relevant information resulting from non-compliance will be included in the safety file maintained by the principal contractor and passed to the client on completion.

Effectiveness of quality assurance

By including a description of quality assurance applied to civil engineering design and construction in this book, the authors do not wish to imply that the procedure is necessarily safer than the more usual traditional contract system. However, when applied to construction and to design systems, quality assurance methods provide a logical and orderly approach to the tasks in hand. They give experienced staff the opportunity to make their contribution to the work strategy and they ensure that the work is checked by competent people.

Civil engineering contracts employing quality assurance procedures are known to have been successful and safe, but too few have been carried out for any conclusions to be drawn on a statistical basis. Nevertheless, it can be confidently argued that some well-publicised disasters would not have occurred if their design and construction had been quality assured.

7 Safety and reliability

Risk

The imperfect nature of the world in general and of construction in particular may lead to the acceptance of a limited number of accidents as inevitable. In construction a particular type of activity is seldom repeated with an identical set of circumstances and it is therefore practically impossible to follow the routine systems of working or to test finished products in a way that has led other industries to achieve improved safety and reliability.

In previous chapters we have considered ways in which occupational health and safety can be improved during the construction use and maintenance phases in the life of a structure. The subject of safety and reliability however is concerned with the ability of the structure to function without unacceptable risk of loss of life during its design life span.

What is an acceptable level of risk? How safe is safe enough? Are engineering failures always failures by engineers? Engineering involves decision making but the decisions are not always made by engineers because they often involve probability and cost. A simple example is the choice of the height of a sea wall.

Enough is known about tides to enable experts to predict the highest water level that is likely to be experienced in the vicinity in a given number of years. This is referred to as the return period. North Sea oil and gas platforms are designed for a sea state of a 100-year return period. Where a sea wall is designed for a 100-year return period, collapse of the wall, flooding, and possible loss of life, has to be accepted as a possibility once every 100 years. A higher sea wall to reduce the frequency of flooding and loss of life could involve massive extra cost and the justification for this has to be left to the politically elected representatives, local or parliamentary. The risk can never be eradicated but it is possible in many cases to quantify the probability of failure so that decisions may be taken on the financial implications of reducing it.

Principles of risk analysis

Risk analysis is based on probabilities derived from past experience. To illustrate the principle let us use it to assess the priorities of proposals for road improvement schemes. For instance, at a gap in the central reservation of a dual carriageway there have been 25 serious-injury accidents in the previous five years, during which time 20 000 vehicles have passed the spot every day. The number of vehicles that have passed the spot in the past five years is $20\ 000 \times 365 \times 5 = 3.65 \times 10^7$ and therefore the serious accident rate is $25/3.65 \times 10^7 = 6.85 \times 10^{-7}$ per vehicle.

But from statistics collected by the Department of Transport, it is not unreasonable to expect that the accident rate can be reduced by 25% by the elimination of right turns on such a dual carriageway. Therefore, all other things being equal, closing the gaps in the central reservation can be expected to reduce the serious accident rate to $0.75 \times 6.85 \times 10^{-7} = 5.14 \times 10^{-7}$ per vehicle.

The estimated cost of the road improvements to eliminate the right turns may then be compared with savings in the cost of the accidents. By calculating other accident blackspots in a similar way, a table of priorities for improvement schemes could be drawn up to distribute a given budget in the most cost-effective manner.

The prediction of the breakdown of a piece of equipment or, conversely, its reliability, can be expressed in the same way. From records of past breakdown, the probability that the equipment will break down in any given period may be interpolated and, with the use of subjective judgement and quantitative mathematics, an estimation of the risk of an accident to which that particular equipment breakdown played a part can be derived. Fault-tree analysis and failure modes and effect analysis enable clearer understanding of how the breakdown of one piece of equipment affects the probability of a major hazard.

Fault-tree analysis

A fault tree is a graphic representation of the sequence of events that could cause the hazard under consideration. Figure 7.1 is a simple fault tree. The hazard, known as the top event, in this case is an explosion in a pumping station. The explosion will only occur where there is a methane/air mixture within explosive limits AND there is ignition. Ignition can occur from electrical equipment OR someone smoking OR from hot surfaces. Similarly, electrical equipment may be faulty OR incorrectly designed. Methane within explosive limits requires oxygen AND methane build-up, which may occur from methane seepage from the soil AND insufficient ventilation.

Over the years a fairly consistent set of fault-tree symbols has been established, some of which are shown in Fig. 7.2. The most important of the event symbols are the circle representing a basic event and the rectangle representing a fault from which further branches of the fault tree stem through the AND and OR gates.

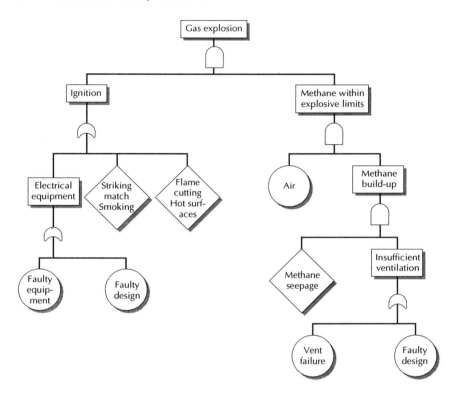

Fig. 7.1.   Simple fault tree

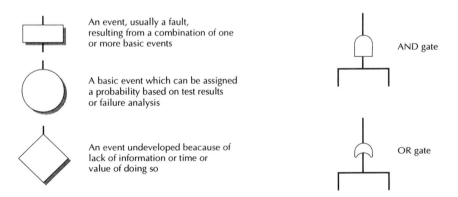

An event, usually a fault, resulting from a combination of one or more basic events

A basic event which can be assigned a probability based on test results or failure analysis

An event undeveloped beacause of lack of information or time or value of doing so

AND gate

OR gate

Fig. 7.2.   Some fault-tree symbols

A fault tree requires that the event is described either by success or failure. Thus, although a control valve in a pipe might seize up in any position, it must be regarded for the fault tree as being either sufficiently open or insufficiently open. Software packages are available to build up fault trees by computer graphics.

Failure mode and effect analysis

Failure mode and effect analysis is the study of potential failures in any part of a system to determine the probable effect on the other parts of the system and on operational success. It is a mental process that every designer goes through but improved by the discipline of recording the analysis in a logical and checkable manner and of identifying the severity of the risks.

The function of failure mode and effect analysis are

- to identify known and potential failure modes
- to identify the causes and effects of each failure mode
- to give each identified failure mode a priority number according to the probability and severity of its risk and the chance of detection before the failure occurs
- to provide for corrective action.

Table 7.1 is an example of failure mode and effect analysis. The risk priority number is a product of the estimates of occurrence, severity and detection, and provides a relative priority of the failure mode. The higher the number, the more serious the failure mode. The occurrence of failure is evaluated on a scale of 1 to 10, 1 indicating a very low probability of occurrence and

Table 7.1. Failure mode and effect analysis for support bracket of a water circulating pump

Failure mode	Cause of failure	1*	2*	3*	$N^\dagger$
Bracket bends	Design fault	2	5	8	80
	Construction error	4	5	8	160
Bracket corrodes	Inadequate protection	3	5	8	120
	Leak	4	5	8	160
Bracket fails	Design fault	1	8	5	40
HD bolts loosen	Incorrect torque	5	5	8	200
	Incorrect nut or bolt	4	5	8	160
HD bolts fail	Construction error	4	8	5	160
	Design fault	2	8	5	80

* 1. Occurence: 1 for low probability, 10 for high probability
  2. Severity: 1 for minor nuisance, 10 for serious hazard.
  3. Chance of detection: 1 for high probability, 10 for low probability.
$\dagger$ N Risk priority number = column 1 × column 2 × column 3.

10 indicating a near certainty of occurrence. The severity of failure is judged on a scale of 10, 1 being a minor nuisance and 10 a very serious consequence. The detection of failure is also judged on a scale of 10, 1 indicating a high probability that failure would be detected before the item is put into service and 10 showing a very low probability of detection. In Table 7.1 the 'bracket fails' failure mode of the tie bar bracket is rated at $1 \times 8 \times 5 = 40$ whereas the 'HD bolts loosen' failure is rated at $5 \times 5 \times 8 = 200$, indicating that tightening the fixing bolts is more of a potential problem.[1]

Reliability

The techniques used for safety and reliability analysis are not solely concerned with the prevention of accidents. They also have an important economic value, for they can be utilised to evaluate the financial consequences of equipment failure and disrupted production. Analysis of the probability and the costs of corrective modifications help management to make cost-effective decisions about preventive maintenance repairs and replacement of plant.

Safety cases

Risk analysis, fault trees and failure mode and effect analysis are also used in the preparation of safety cases. The Flixborough explosion of 1974 (Fig. 7.3) focused attention in the UK on the problems of major hazards. Other major industrial accidents in Europe stimulated the European Community to prepare in 1977 a directive on major accident hazards. This directive is implemented in the UK by the Control of Industrial Major Accident Hazards (CIMAH) Regulations 1984.

A manufacturer who has control of an industrial activity defined in the CIMAH Regulations is required, among other things to submit to the Major Hazards Assessment Unit of the Health and Safety Executive (*see* p. 43 and Fig. 3.1) a written report known as a safety case, which demonstrates that the hazards and risks from the industrial activity have been identified and that the appropriate safeguards have been analysed and found acceptable. The safety case

- identifies the nature and scale of use of dangerous substances
- places the site in its geographical and social context
- identifies the type, relative likelihood, and consequences of potential major accidents
- identifies the control regimes and systems for the site
- identifies whether the manufacturer or occupier of the site has considered the adequacy of the controls.

Whereas much of the preparation of safety cases is carried out by mechanical, electrical and chemical engineers, it is desirable that civil engineers understand the principles, partly because of the civil and structural engineering elements of industrial and nuclear sites, and partly because of the

Fig. 7.3.   Flixborough disaster, 1974

relevance to planning and site investigation for new projects that fall within the regulations.

The Hazards Forum

The recognition of the obligation of engineers to safeguard the public interest in matters of health and safety has been demonstrated at public enquiries investigating several major disasters during recent decades including Flixborough, Herald of Free Enterprise, Hillsborough Football Stadium, Kings Cross, Clapham Junction, Piper Alpha and the collapse of box girder bridges at Milford Haven and West Gate in Australia. This recognition applies not only to man-made disasters but also to natural disasters such as the Towyn Flood where the government needs to be advised on the potential for damage and the precautions to reduce its severity as well as the extent of planning for disaster relief.

The Hazards Forum, set up by many professional engineering institutions and some scientific bodies 'exists to provide a national focal point in which engineering features in the mitigation and reduction of both man-made and natural hazards and disasters'. It aims to make professional engineers more aware of their moral responsibility for health and safety, it co-ordinates initiatives and serves as a platform for detailed discussion of the many scientific and sociological aspects of safety as well as cost benefit analysis.

The Forum organises lectures, seminars and conferences and publishes a quarterly newsletter. The Hazards Forum secretariat is at 1 Great George Street, London SW1P 3AA.

Standing Committee on Structural Safety

The Standing Committee on Structural Safety is an independent body established by the Institutions of Civil and Structural Engineers and others in 1976 to maintain a continuing review of building and civil engineering matters affecting the safety of structures. The prime function of the Committee is to identify in advance those trends and developments which might contribute to an increasing risk of structural unsafety. To that end, the Committee interacts with the professions, industry and government on all matters concerned with design, construction and use of building and civil engineering structures.

The Committee considers current practice, trends and innovations in design, construction and maintenance and considers whether unacceptable risks exist and whether research and development appears desirable from a structural safety standpoint. To assist in meeting its objectives, the Committee seeks feedback from those engaged in construction who, by dint of personal experience may obtain warnings of potentially adverse structural developments. 'Near-misses' in structural adequacy may well occasionally occur as a result of oversight, misconception or hazard which might not have, but could have been anticipated, and should be anticipated in the future. If such is the case then a prompt consideration of potential hazard or weakness is in the public interest.

The secretariat of the Standing Committee on Structural Safety is at 11 Upper Belgrave Street, London SW1X 8BH.

8 Information

The need for information on safety

The requirement of section 2(2)(c) of the HSW Act that 'employers are to provide such information as is necessary to ensure the health and safety at work of his employees' is a very onerous one in construction. The industry is so diverse that almost every type of accident and environmental health hazard is possible for its employees. To ensure that information reaches those who face such a multitude of hazards requires good organisation and dedication.

It is fairly evident that information is necessary to advise engineers, technicians, architects and all those who are in a position to adversely affect the health of construction workers and members of the public by their acts or omissions. Many of the obvious hazards are not covered in university degree coursework or in textbooks and means have to be found to provide professionals with the knowledge they need to avoid repetition of problems experienced by others.

Information is a prerequisite for the communication of the art of good safety practice to all employees. It may be either advisory or mandatory. Advisory information is available in publications, posters and films. However, information sheets or practice notes and mandatory procedures will invariably have to be prepared in-house to give proper emphasis to the particular company's procedures and requirements.

Information prepared in-house

The primary medium for making safety information available is print. It takes the form of handbooks, standing instructions, practice notes and advisory notes prepared for or by the company safety adviser for distribution to those for whom it is intended.

Organisation for preparation and circulation

In most companies within the construction industry there are many subjects – health and safety is only one – that require the circulation of information to employees. It is common practice for the person responsible for administration to establish a routine for achieving this.

A system that operates in many companies is that the authors of instructions or practice notes send the completed and approved texts to the administration department for printing and circulation. While many small firms rely on less formal systems for circulating essential information, the principle that information about health and safety matters should reach those for whom it is intended by design, rather than by chance, remains the same.

Safety handbooks

To ensure that all staff and operatives whether employed by contractors, client organisations or professional firms are aware of the hazardous nature of the industry, it is recommended that a company safety handbook is produced and given to employees when they are first appointed. Some companies include chapters on safety and welfare within a general staff handbook but a separate publication is preferable. However, there are some procedural matters that are common both to safety and general management, for example accident reporting, and it is important for the staff handbook and the safety handbook to be compatible.

The purpose of a company safety handbook is three-fold. First, it sets down briefly and simply the responsibilities of the employer under the HSW Act. Second, it tells the employees what their duties are and, third, it outlines the hazards of the job and how they should be avoided. Because of the diverse nature of some larger companies this third function of describing hazards is contained in a separate booklet or in guidance leaflets aimed at the relevant section of the workforce.

A typical safety handbook

Safety handbooks may contain a synopsis of the company's health and safety policy statement and a personal message from the chairman (*see* pp. 174 and 176). The following is an example of a typical safety handbook for a firm of civil engineers where the policy statement is published elsewhere.

General Introduction

The Health and Safety at Work etc. Act 1974 requires employees to safeguard their own health and safety and to co-operate with their employers in ensuring that the well-being of other company employees and the public at large is not put at risk.

This booklet outlines the duties under the Act that govern the responsibilities of the employer and employee and provides some pointers to safety in circumstances commonly encountered in our business. Please read it carefully and keep it in a safe place for reference purposes.

While the Act is directed only to Great Britain (another similar Act applies to Northern Ireland), its principles should be taken to apply wherever company personnel are operating.

Employer's Responsibilities

The duties of employers in complying with the Health and Safety at Work etc. Act 1974 are outlined in the company health and safety policy statement that is attached to principal notice boards. The statement also describes the company policy on health and safety together with the organisation set up to implement it. The Act places demands on both employer and employee in order to achieve its principal aims of

- securing the health and safety of people at work, and
- protecting others from hazards arising from work activities.

The company, for its part, will

- provide you with a safe working place commensurate with the nature of your job
- inform you of any hazards connected with your job and train you to safeguard against them
- where the risks associated with your workplace cannot be reduced to an acceptable level, provide you with personal protective equipment
- maintain procedures and systems of work to assist you in carrying out your job with minimum risk of damaging the health of others
- support any reasonable contentions you may make that the conditions of any location in which you are required to work are hazardous beyond the nature of your job.

Employee's Responsibilities

For your part you are required to

- take notice of health and safety notices and literature
- know your fire drill: obey fire instructions and office evacuation procedures
- know what to do in the event of an accident and, where called on, do it
- wear clothing and footwear suitable for your job
- wear protective clothing and use safety equipment where your task is hazardous or if you have been instructed to do so
- ensure that, by your acts or omissions, you do not put at risk the health and safety of your fellow employees, other people at work or anyone else
- avoid misusing anything provided for your safety
- attend the training courses provided for you
- obey the health and safety instructions in force at any site or other establishments you may visit in the course of your employment with the company
- be mentally and physically fit for the work you have to do.

General safety information for employees

It is usual for the company safety handbook to contain examples of the principal forms of accident in the construction industry, some of the environmental hazards relevant to employees of the company and the precautions to be taken. Another section usually contains details of the protective clothing and safety equipment available to employees, information and training, and first aid and reporting procedures in the event of an accident.

Organisations that have specific safety requirements such as some industrial undertakings, utility companies, railways and water companies require their employees, and anyone else on their premises to obey a number of safety rules. These need to be included within the safety handbook or, better still, as a separate booklet that may be issued to all contractors and consultants who work for them as well as to their own employees. (Note the *HMS Glasgow* case at p. 29.)

Company safety literature for managers

Managerial instructions vary from company to company. They may be referred to as standing instructions, practice notes, procedural notes or management memoranda. There is a requirement for several grades of staff to receive company instructions on their duties in regard to health, safety and welfare. Managers are not only employees but they also represent the employer to all staff and operatives who report to them. It is therefore necessary to ensure that they are thoroughly aware of the duties imposed on them by law and company policy. Staff who require instructions on health and safety include divisional and departmental managers employed by authorities, boards and major companies, construction managers and site managers employed by contractors and project managers and resident engineers employed by consulting engineers. The subject matter of the managerial instructions would include the terms of reference and duties described in chapter 5.

Practice notes. Most authorities and companies involved in construction activities find it necessary to prepare and circulate information sheets on procedural or technical matters. Examples of procedural matters that might be covered on information sheets include arrangements for safety committees and safety representatives; arrangements for protective clothing and equipment; accident reporting procedures; permit-to-work systems and lone working.

Information sheets and guidance notes on most construction hazards are already available in printed form from the Health and Safety Executive and other sources. However, they are often too general for distribution to all employees, and companies need to explain in-house procedures, giving necessary information to ensure that decisions are made at the right managerial level. Thus it is necessary for companies to prepare practice notes describing hazards that particular employees have to face and the precautions to be taken. A few examples of such practice notes that have been prepared by various contractors for their own use include those for

- working in sewers and confined spaces
- safety precautions during prestressing operations
- safety precautions when working on live highways
- working near live underground and overhead services
- precautions when surveying old and unoccupied buildings.

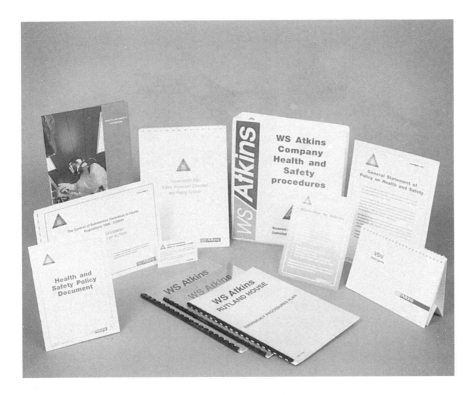

Fig. 8.1. Examples of instructions and guidance notes (courtesy W.S. Atkins Consultants)

An organisation must be set up within the company not only to distribute the practice notes but also to ensure the regular updating both of the contents and the circulation lists. Examples of instructions, practice notes and a safety handbook prepared in-house are shown in Fig. 8.1.

Posters

A number of subjects are more suited for display on notice boards than for circulation to individual staff members. These include local arrangements for health and safety, including the names of safety representatives, first-aid officers and the procedures for reporting accidents.

Fire and evacuation procedures and particular hazard warnings are other subjects suitable for posters which may be either specially produced or obtained from the Royal Society for the Prevention of Accidents (RoSPA) and the British Safety Council. The range of the posters is continually extended, some subjects are updated or deleted and catalogues of currently available posters may be obtained from the organisations concerned. Examples of posters are given in Fig. 8.2.

Fig. 8.2. Examples of safety posters available from RoSPA

Publications

Some publications specialising in safety and health are suitable for wide circulation throughout the staff membership, while others are suitable for reference, for lending libraries or for retention by managers and safety advisers. The specialism and size of the company will determine the way in which safety information is made available to staff.

Publications produced by the Health and Safety Executive

A comprehensive range of free and priced publications is produced by the Health and Safety Executive. Advice about the publications may be obtained from HSE Information Centre, Broad Lane, Sheffield S3 7HQ, Tel 01142 892345. HSELINE is a computer database of bibliographic references to published documents on health and safety at work. It contains over 150 000 references and over 12 000 additions are made each year. For information on how to access HSELINE contact the British Library Science, Reference and Information Service, IRS/Dialtech, 25 Southampton Buildings, London WC2A 1AW, Tel 0171 323 7946.

A list of Health and Safety Executive publications relevant to the construction industry is given in Appendix 4.

Construction safety manual

A manual published by Construction Industry Publications Limited in Birmingham entitled *Construction safety* is sponsored by the Building Employers Confederation (BEC), the Federation of Civil Engineering Contractors and also the Construction Industry Training Board. Other bodies such as the Health and Safety Executive and the Department of the Environment have also collaborated in its production. *Construction safety* is regarded as the standard reference work and training manual for those in the construction industry. It contains over 30 chapters ranging from legal and organisational matters to specific safety topics and is published in two volumes in loose-leaf format so that updated pages can easily be substituted. It is necessary for employers to ensure that their copies of *Construction safety* are kept up to date and those with several copies are advised to give them catalogue numbers, and arrange for their whereabouts to be recorded by a librarian or administrator, who will be able to call the sets in for revision when new or revised pages are issued by the publishers.

Other publications available from Construction Industry Publications Limited on health and safety include

- *Safety policies for construction*
- *CDM Regulations*
- *Safety advice in construction*
- *Risk assessment in construction*
- *COSHH in construction*
- *Noise in construction*
- *Work experience in construction*

Safety manual for mechanical plant construction

The *Safety manual for mechanical plant construction* is published by the Oil and Chemical Plant Constructors Association, and is a loose-leaf publication in two volumes. It is suitable for engineers involved in the petro-chemical industry.

CIRIA site safety handbook

The Construction Industry Research and Information Association (CIRIA) publishes *Site safety: a handbook for young construction professionals* to alert young professionals to the more common hazards present on site, to advise them on safer construction practices and to help them manage their responsibilities. *Site safety* is available from CIRIA, 6 Storey's Gate, London SW1P 3AU. *See also* details of the CIRIA video *Not just an accident . . .* (p. 240).

Books on health, safety and welfare

Books on health, safety and welfare useful to those in the construction industry include

- *RoSPA health and safety practice* by J. Stranks and M. Dewis, Pitman
- *Health and safety* by J. Hendy and M. Ford, Butterworth (supplements by subscription)
- *Croner's guide to health and safety,* Croner
- *Health and safety at work* by M. Goodman, Sweet and Maxwell
- *Opportunities and impositions* by D. Bishop, Construction Industry Council.

Magazines and journals

Staff and particularly middle managers responsible for safety are encouraged to read one or more of the journals produced by safety organisations. Members of the Institution of Occupational Safety and Health receive *The safety and health practitioner*, which is also available to non-members. Members of RoSPA receive monthly *The RoSPA bulletin* and *Occupational safety and health*, also available to non-members. Subscribers to the British Safety Council receive *Safety management* monthly. *Health and safety at work* is a widely read monthly magazine published by Tolley. Circulation of the journals to a wide readership among management and staff is a good cost-effective way of improving their safety awareness.

For safety advisers and others with managerial responsibility *Construction safety information service* is available by subscription from Construction Industry Publications Limited which supplies abstracts of articles and other matters relating to construction safety. Those interested in safety and reliability may wish to consult the *Safety and Reliability Society Journal*. *Site safe news* published by the HSE twice a year is suitable for a wide circulation on construction sites. It is available free from Sir Robert Jones Memorial Workshops, Unit 3 Grain Industrial Estate Liverpool L8 4XY. Items of interest for inclusion in *Site safe news* should be addressed to The Editor, Site Safe News, HSE, Bootle, Merseyside L20 3TW.

Films and videos

The CIRIA video *Not just an accident* which runs for 30 minutes alerts young professionals to the risks they face on construction sites and it displays the need for adequate health and safety planning and management. The video was made in co-operation with the Safety Panel of the Institution of Civil Engineers which identified a need for a video that approached the subject in a manner suitable for graduates and students. The messages to young professionals are: be aware of the risks; don't allow unsafe practices; and plan and manage for health and safety. *Not just an accident* is available from CIRIA, 6 Storey's Gate, London SW1P 3AU.

Videos on construction safety and related subjects are also available from CFL Vision, PO Box 35, Wetherby, Yorkshire LS23 7EX, who also distribute videos made by the Health and Safety Executive, and Longman Training, Longman House, Burnt Hill, Harlow, Essex CM20 2JE. The videos available are too numerous to mention here but those that have been found useful for the general training of young professionals are

- *No questions asked*
- *Eyes down*
- *One step away*
- *Watch that space – confined spaces in the construction industry*
- *Scaffolding safety checkmate I*
- *Scaffolding safety checkmate II*
- *The contract*
- *Falsework*
- *Lift off*
- *A site safer*
- *Hangman (falls)*
- *Beginners luck*
- *Construction safety: the cost of chaos*
- *Construction safety: excavations*
- *Construction safety: scaffolding*
- *Construction safety: lifting equipment*
- *Construction safety: mechanical aids*
- *A guide to manual handling assessments*

The Construction Industry Council (CIC) video *A safer bet – an introduction to the principles of the CDM Regulations* heightens awareness of the contribution designers can make to the health and safety of those in construction. It runs for 35 minutes and is available from CIC Publications, 26 Store Street, London WC1E 7BT.

Many large organisations make in-house films and videos for staff safety training and for contractors who work on their establishments. Some films and videos on the more popular safety topics such as action in the event of a fire are available from public libraries and also some excellent fire films are available on free loan for training purposes. Enquiries may be made to the Fire Protection Association in London, or to any county fire brigade.

9 Training

The need for safety training

The general duty of the employer to 'provide such instruction and training as is necessary to ensure the health and safety at work of his employees' and, presumably, others who may be affected by the acts or omissions of his employees, leaves little doubt that the first questions to be asked at an enquiry after an accident are 'Were the staff and operatives competent? Have they been properly trained?' To satisfy the general duty of safety training, employers in the construction industry find it necessary to provide training at several levels from a basic introduction to health and safety for everyone in their employment to safety training courses for managerial staff. Moreover, safety should form part of all training of construction operatives.

Induction courses

Many companies in the construction industry provide induction courses for all new employees. The courses include an outline of the organisation and management of the company, personnel and welfare arrangements and staff facilities available. The opportunity must not be missed to include a talk on health and safety. Points that must be stressed include

- the necessity of reading the company safety handbook
- that the need to heed personal safety and the safety of colleagues and the public at large is a requirement not only of the law but of employment with the company; breach of that duty can result in a reprimand or, in extreme cases, dismissal
- how to obtain first aid
- what to do in the event of a fire
- what to do in the event of an accident
- how to evacuate the building
- how to obtain protective clothing and safety equipment
- a brief outline of the likely hazards employees may encounter, together with examples of accidents that have occurred

- a brief outline on further safety training and information that is required for managers, supervisors, designers and site personnel.

It is recommended that the basic introduction to health and safety should last for at least two hours and should include visual aids, slides and possibly a film.

Graduates and technicians

It is recommended that time is devoted on the relevant courses at universities or colleges of technology to the subject of health and safety in the construction industry. The training should cover an outline of safety legislation and the duties of employers and employees. It should describe the hazards that professional staff face in the course of their work and outline the problems facing designers in ensuring that their work is safe to build, operate and maintain. Emphasis should be given to the practical ways in which graduates can help to improve the poor safety record of the construction industry.

The subject must be expanded on during the postgraduate training period. All civil engineering graduates and other professionals should have formal lectures in health and safety legislation, site safety and safe systems of working. At their professional examinations graduates and technicians are expected to satisfy the examiners regarding their knowledge and experience in basic health, safety and welfare practice.

From 1996 all candidates for the Chartered Professional Review of the Institution of Civil Engineers must have had three days specific training on health and safety matters as part of their continuing education and training. The three days cover

- familiarity with health and safety legislation
- appreciation of general site health and safety practice
- knowledge of safety in the candidate's specific work area.

Candidates for the Incorporated Professional Review leading to Associate Member status require two days health and safety training.

At the Chartered Professional Review, candidates are required, among other things, to write an essay of 1½ hours on a topic set from a list published by the Institution. The list contains several questions on health and safety and related matters and most candidates find themselves tested on the knowledge they have gained from their three days of health and safety training.

Managers

It is insufficient for senior staff merely to have attended the basic induction course and graduate training in health and safety. Their duty as representatives of their employer has to be made quite clear to them and it is therefore desirable that all managers attend a two-day or three-day course on management in construction. The syllabus for the course depends, to some extent, on the branch of the industry to which the particular employer belongs.

Items on such a course should cover

- the law: health and safety legislation applicable to the construction industry, and the law of contract and tort with special reference to duties and responsibilities for safety
- safety policies: the policy of the particular employer and methods of implementation, policies of other authorities, contractors and consultants, and the role of the company safety adviser
- hazards: the principal causes of accidents on site, and occupational health hazards
- accident prevention: application of the regulations specific to construction and other relevant regulations and codes
- accidents: procedures in the event of an accident, and accident reporting.

Specific safety training items will need to be added to the syllabus depending on the branch of engineering of the middle managers attending the course. For example, construction and site managers need to cover in greater depth the regulations, notices, records and test certificates that must be complied with. In particular, they need to have a thorough understanding of the CHSW Regulations, the CDM Regulations, the MHSW Regulations and other regulations and codes relevant to their particular line of business, and the ways in which the requirements can be met.

Project directors, partners and managers with professional firms need to understand how the company meets its statutory duty to provide for the safety of its employees especially when away from the office. In addition, they need to cover

- statutory and common law duties relating to designers and specifiers
- company quality procedures and their implementation
- contractual relationships with clients and contractors on safety matters
- company procedures for providing safety information and training for and supervision of its employees.

Managers employed by government, utilities and other major employers need to cover the safety aspects of their own industry and how their particular organisation enforces its safety procedures in relation to its own employees and outside contractors. In this connection, knowledge of the duties of the 'client', the 'planning supervisor' and the 'principal contractor' under the CDM Regulations is vital.

Planning supervisors

The duties of the planning supervisors appointed by clients under the CDM Regulations are sufficiently separate from main-stream project management to warrant additional specialised training.

It is a prerequisite that planning supervisors are already familiar with health and safety law, and with construction hazards and their prevention. In addition, the planning supervisor needs training to cover items which may be less familiar, such as

- judging the competence and resources of designers and contractors
- the 'notification' procedure
- understanding the designers duties and seeing that they are complied with
- preparing a health and safety plan
- keeping a health and safety file.

This additional training for planning supervisors is worthy of a two-day course which can be added to a project managers safety course or taken after practical experience as a project manager.

Larger companies should consider running planning supervisor courses in-house because there are advantages in adopting a house style to many of the duties. Otherwise there are many commercial courses available for training planning supervisors one of which is run by CTA Services.

Safety advisers

Where a safety adviser or safety officer is employed full time it is recommended that he attend a course leading to qualification as a member of the Institution of Occupational Safety and Health. The training programme recommended for construction safety advisers is specially developed by the Building Employers Confederation, the Federation of Civil Engineering Contractors, the Construction Industry Training Board (CITB) and the Institution of Occupational Safety and Health, with the help and co-operation of the Health and Safety Executive. Full details of the scheme of training are given in CITB Training Recommendation 13, and details of the courses may be obtained from Brooklands School of Management in Weybridge, Surrey.

Resident engineers

The particular situation of resident engineers, or Engineer's Representatives under the ICE Conditions of Contract, should be covered by a training course of at least two days with a syllabus drawn from elements of the middle management training syllabus. The resident engineer has to be thoroughly aware of his responsibilities under the HSW Act and how to relate this to his duties under the Contract. The engineer's safety adviser, appointed to assist the resident engineer in day-to-day safety matters, should attend the same training course.

Other training courses

Courses in safety for site management are held regularly at the CITB National Training Centres at Bircham Newton, Glasgow, Erith and Birmingham. The courses, which are recognised by the Institution of Civil Engineers for continuing education and by the Chartered Institute of Builders, are of one week duration, and on successful completion, the Site Manager's Safety Certificate is awarded.

The many courses run by the CITB at their four national training centres include elements of safety. However, there are a number of specialised safety courses for site management in addition to the one previously mentioned.

- Construction Safety Advisers Certificate Course (15 days)
- Safety Officers: Crane (3 days)
- Safety Officers: Scaffolding (3 days)
- Safety Officers: Support to excavations (3 days)
- Safety in Demolition (supervisors and foremen) (1 day)
- Site Management Safety Update (1 day).

Details of these courses may be obtained from CITB Bircham Newton Training Centre, Kings Lynn, Norfolk or from the other national training centres (Figs 9.1 and 9.2).

For safety advisers or managers wishing to gain a basic knowledge of safety in the construction industry a course in safety for middle management may be suitable. The RoSPA Occupational Safety Training Centre in Birmingham run a four-day course to make delegates aware of the responsibilities placed upon both the client and the contractor and to provide an understanding of the hazards and precautions of construction, demolition and maintenance. Several courses of one-day and two-day duration on a range of topics associated with construction are offered by RoSPA. Details may be obtained from RoSPA, 22 Summer Road, Acocks Green, Birmingham B27 7UT.

CTA Services, a division of Thomas Telford Services Ltd, a company wholly owned by the Institution of Civil Engineers, offers training courses at various venues in the UK or by arrangement with companies to meet their specific training requirements. They also operate distance learning in civil engineering law and related subjects. Health and safety topics covered by CTA include comprehensive training for the design team leader and planning supervisor in relation to CDM Regulations (seven days), Role of the Designer on CDM Regulations (two days) and one-day courses on

Fig. 9.1 Training in safe trenching practice at Bircham Newton Training Centre (courtesy CITB)

- health hazards in construction
- responsibilities of resident engineer and architect
- scaffold inspection
- risk assessment in construction
- signing and guarding of roadworks
- the New Roads and Street Works Act 1991.

Details may be obtained from CTA Services, Belmont House, 11/12 Devizes Road, Swindon SN1 4BH.

In addition to those mentioned previously a number of safety training centres in Britain and Northern Ireland cover the needs of companies who

Fig. 9.2 Training plant drivers and banksmen at Bircham Newton Training Centre (courtesy CITB)

do not have sufficient resources to run in-house courses. Details of these may be obtained from the Federation of Civil Engineering Contractors, the Building Employers Confederation or the Health and Safety Executive.

For first-aid training in accordance with the Health and Safety (First Aid) Regulations 1981, courses held by the St John Ambulance, St Andrews Ambulance Association and the British Red Cross Society (*see also* p. 270) are approved by the Health and Safety Executive.

Training in fire prevention and action in the event of a fire may be obtained by application to any county fire brigade or to the Fire Protection Association in London. The fire brigade also give training in confined space entry and rescue, including the wearing and maintenance of breathing apparatus.

Training records

The training of operatives and staff at all levels and their competence to do the work required of them is a legal duty imposed on employers under section 2 of the HSW Act and in most regulations made under it. It is unlikely that anyone but a few employees in the industry can work legally without some sort of safety training.

Because the provision of training is a legal duty and therefore may, and often does become an issue in the courts, it is important that employers keep accurate records of all training received by their employees and others, so that evidence can be produced if necessary. The records are also useful for identifying absentees and for re-training purposes.

10 Protective clothing and safety equipment

General requirements

The wearing of protective clothing and the use of safety equipment is crucial in the battle to reduce the effects of accidents on construction sites. Employers have a legal duty to provide protective clothing and equipment free of charge whenever other methods to control risk to health and safety cannot provide sufficient safeguards. This would apply to the construction industry most of the time. Employees also have a duty to preserve their own health and safety and it is in everyone's interest that they should be adequately clothed whether special clothing is provided by the employer or not.

Disciplinary action may be taken against employees who refuse to wear the protective clothing supplied by their employers. The refusal to wear protective clothing was decided by the court to be a dismissable offence in a specific case: *Marsh* v. *Judge International Housewares Ltd* [1976].

However, where letters of appointment clearly state that the wearing of protective clothing and use of safety equipment as instructed is a condition of employment and that any refusal to do so is subject to reprimand or in extreme cases dismissal, employees are left in no doubt about their obligations. Employees should also be informed that they have a legal duty not intentionally to interfere with or misuse anything provided for their safety.

Legislation

The MHSW Regulations (*see* p. 194) require employers to identify and assess the risks to health and safety present in the workplace and to reduce those risks to an acceptable level. The use of personal protective equipment to reduce those risks must be regarded as a last resort because it

- protects only the person wearing it whereas other methods of control can protect anyone in the workplace

- is seldom fully effective
- may restrict the wearer by limiting mobility or visibility.

However, if risks cannot be adequately controlled by other means personal protective equipment must be provided and a number of regulations complied with. The regulations commonly associated with personal protective equipment on construction sites are the

- Personal Protective Equipment at Work (PPE) Regulations 1992
- Control of Lead at Work Regulation 1980 (*see* p. 162)
- Ionising Radiations Regulations 1985 (*see* p. 165)
- Control of Asbestos at Work Regulations 1987 (*see* p. 161)
- Control of Substances Hazardous to Health (COSHH) Regulations 1988 (*see* p. 158)
- Noise at Work Regulations 1989 (*see* p. 165)
- Construction (Head Protection) Regulations 1989 (*see* p. 20).

Personal Protective Equipment at Work (PPE) Regulations 1992

These regulations do not generally apply to activities covered by the other regulations listed in the previous paragraph except that, where more than one item of personal protective equipment is worn, they must be compatible with one another.

The PPE Regulations require employers and the self-employed to be provided with personal protective equipment where the risks cannot be adequately controlled by other means. The personal protective equipment must

- be appropriate for the risk and the conditions
- take account of the wearer
- fit the wearer
- be effective for the risk involved without increasing the overall risk
- comply with appropriate design and manufacture legislation.

Employers have to see that personal protective equipment is properly maintained and that it is repaired or replaced when damaged, that it is properly stored when not in use and that information, instruction and training in its use and care are given. It should be noted that employers must ensure employees use the personal protective equipment issued to them and the employees themselves must use it properly and return it to store afterwards. Further, under section 9 of HSW Act (*see* p. 17) no charge can be levied on employees for the provision of personal protective equipment.

Standards of personal protective equipment

It should not be assumed that all protective safety wear is satisfactory for all circumstances. An employer or self-employed person has to make an assessment of the risks at the workplace and the parts of the body in danger. The choice of equipment must then take into account

- the physical effort required to do the job
- the methods of work
- how long the personal protective equipment needs to be worn
- the requirements for visibility and communication.

All protective equipment must comply with UK legislation. The Personal Protective Equipment (Safety) Regulations 1992, which implement the PPE Product Directive, require that almost all protective equipment supplied for use must be certified by an independent body and carry a 'CE' mark. Having made their assessments, employers can then select suitable clothing or safety equipment from the range of 'CE' marked equipment available. If further information is needed to ensure that personal protective equipment provides the protection required, manufacturers or suppliers should be consulted as they have a duty under the PPE (Safety) Regulations to provide such information.

Useful guidance on the selection, use and maintenance of personal protective equipment is to be found in part 2 of *Personal protective equipment at work, guidance on the PPE Regulations 1992* (*see* Appendix 4).

The wearing of protective clothing by professional staff

Whatever the legal requirements, it is essential that professional staff and other managers in the construction industry always wear safety helmets, protective footwear and other clothing, both for their own protection and as an example to the workforce. Visitors and those on short-term or *ad hoc* assignments are prone to ignore safety protection, often because of lack of managerial control, but they are equally vulnerable and should not enter hazardous workplaces unless they are properly protected.

Protective clothing for use on construction sites

The range of available protective clothing and equipment is wide and there are many manufacturers and retailers. We are concerned mainly with the protective clothing and safety equipment used in the construction industry and in particular with that commonly worn and used by civil engineers and other professionals in the course of their site duties. The engineer illustrated in Fig. 10.1 is wearing a standard site safety helmet, parka jacket and safety wellington boots.

Several contractors and major authorities have arrangements with retailers which enable employees to select clothing from a number of styles and sizes. For items of workwear not supplied under the PPE Regulations, purchase at a discount is often available. Some suppliers have mobile showrooms that visit customers on a regular basis.

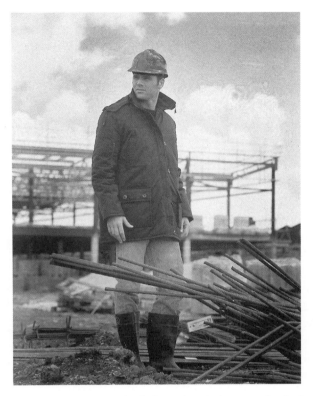

Fig. 10.1. Engineer wearing standard site safety helmet, parka jacket and safety wellington boots (courtesy ARCO)

Safety helmets

Legislation to enforce the wearing of safety helmets on construction sites is embodied in the Construction (Head Protection) Regulations 1989 (*see* p. 18). Safety helmets should be supplied and worn unless circumstances are specifically exempt from the regulations.

Safety helmets must comply with BS 5240: 1987 *Industrial safety helmets*, to be replaced by BS EN 397. To fit, a helmet should be of an appropriate shell size for the person wearing it and have an easily adjustable headband, nape and chin strap. The range of adjustment should be appropriate for the wearer.

Head protection should be as comfortable as possible. Comfort is improved by

- a flexible headband contoured to fit the forehead
- an absorbent, easily cleanable or replaceable sweatband
- textile cradle straps
- chin straps, when fitted, which fit around the ears and otherwise follow the recommendations in the guidance on the Construction (Head Protection) Regulations (*see* Appendix 4).

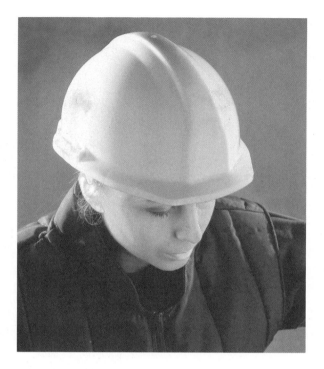

Fig. 10.2.   Reduced peak safety helmet (courtesy ARCO)

Wherever possible the head protection should not hinder the work being done. A helmet with little or no peak may be suitable for a surveyor or a scaffolder. Flexible headbands and Y-shaped chin straps are required for those who have to bend or look up or who work in windy conditions especially at heights. A reduced peak safety helmet is illustrated in Fig. 10.2.

Safety footwear

All workers, including those in offices, should wear sensible shoes or boots to prevent abrasions and cuts from sharp objects on the ground. Construction sites are especially hazardous. Nails, reinforcing bars, binding wire and other sharp objects on rough, uneven and untidy ground result in thousands of foot accidents, some of which become very serious. Provision of foot protection for construction is covered in the guidance on the PPE Regulations (*see* Appendix 4). Recommended safety wellington boots for wet or muddy conditions have a steel midsole, to resist penetration from upstanding sharp objects, as well as a steel toecap. They are shown in Fig. 10.3(a) and are distinguished from the ordinary heavy-duty wellington boot by a red-coloured sole.

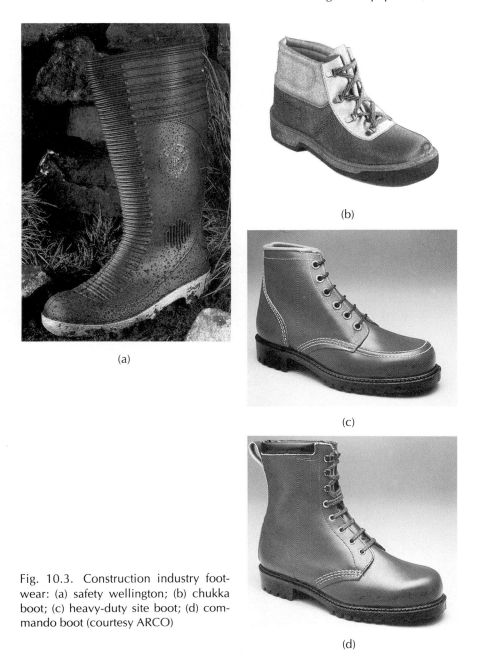

(a)

(b)

(c)

Fig. 10.3. Construction industry footwear: (a) safety wellington; (b) chukka boot; (c) heavy-duty site boot; (d) commando boot (courtesy ARCO)

(d)

The heavy-duty site boots with steel midsoles illustrated in Fig. 10.3(c) are recommended for general site wear. Commando boots and chukka boots are somewhat lighter (Fig. 10.3(b) and (d)) and are suitable for use where the risk of sole penetration is less, but not on construction sites.

Hearing protection

The noise levels in some areas on construction sites are often well above the level which will cause permanent hearing damage to workers in the vicinity. Many people are in fact unaware that their hearing has already been damaged as a result of work or leisure activities and are therefore less susceptible to loud noise and believe hearing protection is unnecessary for them (*see also* p. 164).[1]

The common form of hearing protection in the industry is the ear defender consisting of a headband and cup. There are several types of headband depending on helmet attachment and hairstyle. More important, however, is the choice of cup, which must be selected for the appropriate noise level recommended by the manufacturer. Choice of an incorrect protection will not reduce the sound level by the required amount. Helmet muffs are also available (Fig. 10.4). These can be snapped out into the rest position or swung up on to the helmet when not in use. Ear plugs are also widely used in the construction industry.

Eye protection

The dangers from flying particles and dust are obvious to most construction workers but the wearing of safety eyewear is far from universal and many thousands of painful eye injuries occur on construction sites every year. A number of processes common in the construction industry that cause flying fragments or particles, chemical splash, intensive glare, etc.

Fig. 10.4.   Ear defenders attached to site safety helmet (courtesy ARCO)

are those for which eye protection is required under the PPE Regulations. The regulations apply to all those whose eyes may be injured, e.g. supervisors and inspectors, as well as the operatives.

Goggles, face shields and spectacles are available giving protection against impact, dust, chemical, molten metal, and gas hazards, and managers should ensure that the correct protection and its correct grade is used by the employee (*see* Fig. 10.5). See also BS 7028: 1988 *Guide for the selection, use and maintenance of eye-protection for industrial and other uses.* Eye protectors conforming to BS 2092: 1987 *Specification for eye protectors for industrial and non-industrial use*[2] (to be replaced by BS EN 166, 167 and 168) are marked to indicate the protection they are certified to give. Inspectors of welding need to wear welding goggles with lenses approved to BS 1542: 1982 *Specification for equipment for eye, face and neck protection against non-ionising radiation arising during welding and similar operations* and BS 679 *Specification for filters, cover lenses and backing lenses for use during welding and similar operations*[3,4]

Respiratory protection

The choice of protective equipment against the effects of dust, mist, fumes and lack of oxygen is wide, and great care should be exercised in its selection if the health of the wearer is not to be seriously damaged. It is convenient to consider, first, protection against contaminants in the atmosphere, and secondly, protection against oxygen deficiency. Reference should be made to the COSHH Regulations approved code of practice (*see* Appendix 4).

Respirators. A number of respirators are available for protection against dusts, vapours and fumes. They have face masks to cover the nose and mouth or nose, mouth and eyes and are fitted with renewable filters

Fig. 10.5.   Eye shield to BS 2092 Grade 2 (courtesy ARCO)

according to the protection required. Powered respirators, usually with a full face mask, have filtered air supplied at positive pressure by motor, powered by rechargeable batteries.

Disposable maintenance-free respirators or face masks that meet the requirements of BS EN 149: 1992 *Specification for filtering half masks to protect against particles*, type 1, type 2 and type 3 are suitable for fine dusts; dust and mists; and dust, mist and fumes, respectively (Fig. 10.6). Many other types of disposable respirator are made, for protection against the effects of spray paint, organic vapour, acid gas, ammonia, etc. and nuisance odours.

All respirators should indicate the protection that they provide to prevent incorrect use. Even though they may comply with the relevant British Standards, they must also be approved by the Health and Safety Executive.

Breathing apparatus. Those entering confined spaces to which it is not practicable to provide an ample supply of fresh air, or other atmospheres deficient in oxygen or dust-laden environments which cannot be dealt with by ventilation, must wear breathing apparatus. Standard, positive pressure, compressed-air breathing apparatus is shown in Fig. 10.7(a). Staff must be properly trained in its use and most fire brigades are prepared to give this training. One other person must always be present as a precaution in case of difficulty. The compressed-air cylinders have varied working durations (e.g. 20–35 minutes), depending on their capacity, and arrangements have to be made to recharge them.

Fig. 10.6.   Nuisance dust mask (courtesy ARCO)

(a)   (b)

Fig. 10.7. (a) Self contained breathing apparatus with 15 minute cylinder; (b) airline breathing apparatus with air system trolley (courtesy ARCO)

A system that gives the operative more manoeuvrability and somewhat greater protection is the airline respirator, which is fed by airline from cylinders mounted on a trolley unit (*see* Fig. 10.7(b)). This incorporates a whistle warning unit and up to 90 m of hose.

Outer wear

The guidance on the PPE Regulations indicates that body protection and gloves are necessary to protect against cold and bad weather in construction and other outdoor work. The CHSW Regulations also require protection against adverse weather having regard to any protective clothing provided.

The parka jacket is popular for general site wear by professional staff in the UK (Fig. 10.8(a)). It is warm, waterproof and strong . Donkey jackets are also popular for supervisors and tradespeople. For those required to work in cold and wet conditions, a range of inclement weather clothing, either two-piece or overalls, is available from many manufacturers.

High-visibility garments

Guidance on the PPE Regulations requires that high-visibility clothing should be worn by workers on roadsides and other areas where it is important to be seen to be safe. This would include construction sites where

(a)   (b)

Fig. 10.8.   (a) Waterproof PVC backed nylon coat; (b) waterproof high-visibility coat
to BS 6629 class A (courtesy ARCO)

workers and staff need to be seen by drivers of vehicles and plant. Garments for use on the highway should conform to BS 6629:1985 *Specification
for optical performance of high-visibility garments and accessories for use on the
highway.*[6]

Chapter 8 of the Department of Transport *Traffic signs manual* (*see also*
p. 122) requires all personnel on or near carriageways to wear high-visibility garments to BS 6629 *Class A or B to Appendix G or better* and on
motorways and other high speed roads to *Class A with sleeves to Appendix G.*

A variety of high-visibility jackets, tabards and belts is available for use
on public highways, railways, airports, industrial premises and construction sites where workmen and staff need to be seen by drivers of vehicles
and plant. Garments for use on the highway should conform to BS 6629:
1985.

Other protective clothing

Many items of clothing and equipment worn by construction workers and
staff are not described in this chapter. They include personal protective
equipment required by the Diving Operations at Work Regulations 1981,

the Electricity at Work Regulations 1989 and the Offshore Installations (Operational Safety and Welfare) Regulations 1976. Further information may be obtained from a catalogue of a specialist supplier.

Other safety equipment

Engineers often have to work at heights or go underground. The hazards of heights and confined spaces can never be entirely eliminated by the normal provision of safe working places so personal protection must be worn. Specialist safety equipment for mountaineering, pot holing and caving is used by some engineers for inspecting or surveying high

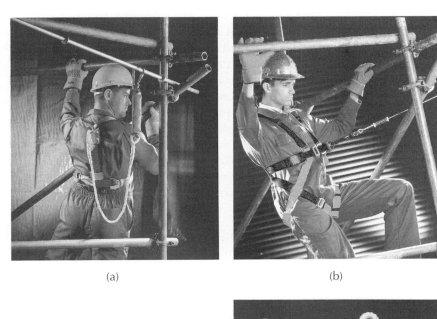

(a)   (b)

Fig. 10.9. (a) Chest safety harness fitted with shoulder straps and waist belt (max fall 2 metres). (b) Full body harness. (c) Retractable lanyards and mechanical rope grab (courtesy ARCO).

(c)

structures or sewers and tunnels and, of course, diving equipment is used for underwater work. However, only the most common equipment is described here.

Safety belts and harnesses

People required to work at a height where the provision of safe working platforms or personal suspension equipment are impracticable should wear a safety belt or harness and line or lanyard attached to a reliable strongpoint above the working position. Full safety harness is preferable to a safety belt because injury is less likely in the event of a fall. The use of fall arrest equipment is covered in the CHSW Regulations.

Harnesses should be fitted with a lanyard, which limits any free fall to 600 mm, or up to 1.2 m provided the lanyard embodies an integral shock absorber. The lanyard for use with a chest or general-purpose harness should allow for a maximum 2 m fall from the point of attachment. Types of safety belt and harnesses are described in BS 1397 *Specification for industrial safety belts, harnesses and safety lanyards*.[7] The use of safety belts and harnesses in the erection of structural frameworks is described on p. 94 and is illustrated in Fig. 10.9(a), (b) and (c).

Chest or body harnesses, not safety belts, should be worn by those entering confined spaces because, when rescue is necessary, an unconscious person needs to be in an upright position to pass through the exit manhole. Lifting an unconscious person wearing a safety belt causes the victim to double over, and rescue through a manhole is then extremely difficult.

11 Accidents and first aid

The accident rate in the construction industry is such that most professional engineers and builders will be involved with an accident at some time in their career. It is therefore in the interests of all engineers, the unfortunate victims and their relatives and friends that professional staff should be fully prepared to deal with accidents when they occur, to provide first aid and to carry out the proper investigation and reporting procedures afterwards.

Site arrangements for health, safety and welfare

Site manager's duties

When work starts on a new construction site, the contractor has to carry out a number of duties that are required by law. In practice, they are usually performed by the site manager.

One of the site manager's first duties is to check that the planning supervisor has notified the Health and Safety Executive under the CDM Regulations (*see* pp. 24 and 196), provided that the project is notifiable i.e. will last longer than 30 days or involve more than 500 person-days on site. Any items on form F10 (rev) that the planning supervisor may not have been able to complete must now be furnished by the principal contractor to the Health and Safety Executive. A copy of Form F10 (rev) or equivalent must be displayed prominently on site by the principal contractor.

The site manager may also request a visit from the Factory Inspectorate to discuss site safety provisions and problems. The Health and Safety Executive should be informed on form OSR1 before occupation of any offices that are to be in use for more than six weeks where fixed (or six months where mobile) where the offices are to be occupied for more than 21 person-hours per week.

It will also be necessary to make an application to the Health and Safety Executive for a fire certificate in cases where offices on construction sites are occupied by more than 20 people on the ground floor or 10 people above the ground floor, or where explosives or flammable materials are stored in or below the premises.

The site manager is recommended to inform the local fire brigade and ambulance service of the location of the new site and its points of access so that in the event of an emergency call for fire or rescue services or for an ambulance the services will be familiar with the site. A meeting may also be requested on site to discuss safety and first-aid provisions.

The site manager must make sure that posters are displayed in prominent places on the site and at site offices, indicating the action to be taken in the event of an injury or dangerous occurrence. A typical site notice is shown in Fig. 11.1.

Constructwell Limited

Site arrangements for health, safety and welfare

Site manager   Tel:

Deputy site manager

Principal contractor

Site safety supervisor

Safety representatives   Tel:

First aiders

The first-aid room is located at

The nearest hospital is

In case of an accident or fire phone

or dial 999 at public phone

The telephone number of the local office of the Health and Safety Executive is

Constructwell Limited:
The director responsible for this site is

The construction manager responsible for this site is

The Company safety adviser is

Fig. 11.1. A typical site notice

Telephones

For reasons of safety at the very least, a site telephone should be installed at the earliest opportunity. The first few days on site are no less hazardous than any other time. Until a telephone line is installed it is recommended that a portable telephone be provided. The emergency services telephone number (999), the emergency numbers of key site personnel and the number of the local office of the Health and Safety Executive should be prominently displayed near the telephone. Portable telephones should always be provided where no fixed telephone line is practicable, e.g. on site investigation contracts, utilities contracts, minor roadworks and surveys.

People allowed onto the site

A duty of the principal contractor under the CDM Regulations is to see that only authorised people are allowed onto the site. This too must be attended to by the site manager early in the contract. For some sites it may be necessary to establish a 'pass' scheme and for others, reporting to security at the main entrance will suffice. Methods for excluding people from the work area are described in HSE guidance note G7 *Accidents to children on construction sites*.

Emergency procedures plan

When the site work includes operations that might be more hazardous than normal a plan should be drawn up by the principal contractor for dealing with any foreseeable major incident. The plan should include methods of evacuation and rescue of all people likely to be on site and should be formulated following discussion with the emergency services (police, fire and ambulance) and other relevant site organisations, e.g. subcontractors, resident engineer and the planning supervisor. The plan is called an emergency procedures plan. Evacuation or rescue or both are particularly difficult where the incident affects people working

- at height
- in confined spaces
- in sewers, particularly live sewers
- in tunnels being constructed where the length of blind heading is over 500 m
- over or adjacent to water
- in remote areas with difficult access.

Depending on the severity of any incident that is likely to occur, preparation of an emergency procedures plan leads to a number of considerations, such as

- provision of training for the contractor's rescue team
- discussions with police, fire brigade and ambulance service on such matters as

- ° training by the fire brigade in the use of breathing apparatus
- ° traffic control in the event of an accident – access routes and assembly areas
- ° clarification of any special rescue equipment that may be required and who should provide it
- ° familiarisation visits by the emergency services – rescue exercises
- the use of a tally procedure or clock on–clock off system to account for everyone on the site
- involvement of site medical staff or first aiders in any major incident
- how an alarm will be given
- approval of the plan by all concerned, including the emergency services, followed by its circulation and publication in prominent positions.

The emergency procedures plan may be accompanied by an event tree, an example of which is indicated in Fig. 11.2.

First-aid facilities

Legal requirements

The Health and Safety (First Aid) Regulations 1981 require employers to provide or ensure the provision of appropriate equipment and facilities enabling first aid to be rendered to their employees if they are injured or become ill at work. Employers must inform employees of the arrangements that have been made for rendering first aid, including the location of equipment, facilities and personnel. Reference should be made to *First aid at work*, the Health and Safety (First Aid) Regulations 1981 approved code of practice and guidance.[1]

First aid at Work gives guidance on complying with the regulations and suggests criteria that can be applied to assess whether first-aid provisions are adequate. For construction sites they include

- the number of employees at risk
- the size of the site and distribution of the workforce
- shift working
- distance from outside medical services.

Thus, on a typical building site, centralised first-aid facilities would be suitable whereas on a scattered civil engineering project it would be necessary to provide a number of first aid stations so that reasonably rapid access to first aid is available to all employees.

Provision for self-employed people

Under the regulations, a self-employed person must provide equipment, or ensure that equipment is provided, to enable him to render first aid to himself at work.

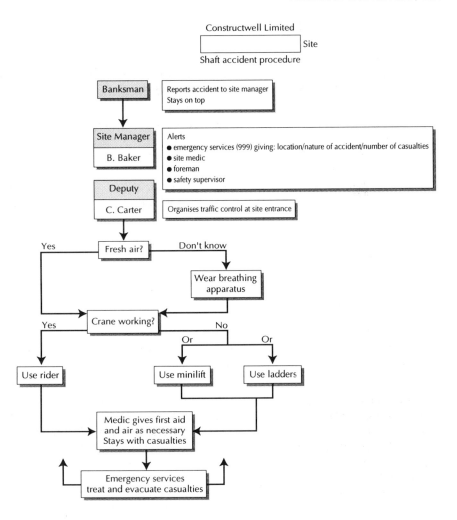

Fig. 11.2. A simple emergency procedures plan (courtesy John Laing Construction Ltd)

First-aid rooms

On all large construction sites and others where access to accident and emergency services is difficult, or where there is dispersed working, a suitably equipped and staffed first-aid room should be provided. This first-aid room should

- be in the charge of a suitably qualified person
- be readily available and used only for first aid
- be of sufficient size and located at a point of access for transport to hospital
- contain suitable facilities and equipment, be ventilated, heated,

lighted, cleaned and maintained, and a waiting room should be provided
- contain a sink with hot and cold running water, soap, nailbrush and paper towels, drinking water and suitable vessels, smooth-topped impermeable working surfaces, a couch with pillow and blankets, clean garments for use by the first aider, an adequate supply of sterile dressings, a clinical thermometer, a store for first-aid materials and a lidded refuse container
- be equipped with telephone, radio and siren for immediate contact with the first-aider on call, and the ambulance and rescue services.

First-aid boxes

Regardless of the number of employees there must be at least one first-aid box on any construction site or work location. The first-aiders must have easy access to first-aid equipment and each box should be placed in an easily accessible location. The boxes should contain nothing but first-aid equipment and be cleaned and checked regularly. First-aid boxes should contain sufficient quantities of suitable first-aid materials and nothing else. In most cases these will be

(a) one guidance card
(b) twenty individually wrapped sterile adhesive dressings
(c) two sterile eye pads
(d) six individually wrapped triangular bandages
(e) six safety pins
(f) six medium-sized individually wrapped sterile unmedicated wound dressings
(g) two large individually wrapped sterile unmedicated wound dressings
(h) three extra-large individually wrapped sterile unmedicated wound dressings.

Where mains tap water is not readily available for eye irrigation, sterile water or normal saline (0.9%) should be kept in sealed disposable containers. Eye baths/eye cups/refillable containers should *not* be used for eye irrigation.

Where employees work singly or in small groups small travelling first-aid kits should be provided and employees should be instructed in basic first aid. The contents of the kit should include at least

- a general guidance card on first aid
- six individually wrapped sterile adhesive dressings
- one large sterile unmedicated dressing
- two triangular bandages
- two safety pins
- individually wrapped moist cleansing wipes.

Shared first-aid facilities

The regulations require employers to provide first-aid facilities for their employees. It follows that other contractors, engineers, professional firms or clients who have staff on site should also make arrangements for the first aid of their employees, especially where their places of work are remote or separate from that of the main contractor. Resident engineers' and subcontractors' site offices should normally have a first-aid kit and trained first-aiders or appointed persons.

Alternatively, subcontractors and others may decide to share first-aid facilities. Two or more employers on a site may make an agreement whereby one of them provides all the necessary first-aid equipment and personnel. The agreement should be recorded in writing and a copy kept by each employer concerned. The employers not providing first aid must satisfy themselves that the employer who has undertaken the duty is carrying it out to a satisfactory standard. Each employer must inform his employees of the first-aid arrangement on site.

First-aiders

Legal requirements for first-aiders

Paragraph 3(2) of the regulations requires an employer to provide an adequate number of suitable people for rendering first aid to employees if they are injured or become ill at work. In order to be suitable for this purpose a person must have undergone

- such training and has such qualifications as the Health and Safety Executive may approve
- such additional training as may be appropriate in the circumstances.

Where the first-aider is temporarily absent it is sufficient for an appointed person to take charge of aid to an injured employee and of the equipment and facilities provided. An appointed person is someone of reasonable intelligence who should be able to handle an emergency and summon help.

For most sites the contractor should ensure that at least one first-aider is normally present with appropriate cover for leave and sickness. On scattered sites and for shift working it will be necessary to provide appropriate cover by first-aiders or appointed persons depending on the degree of risk.

On sites with special risks e.g. sewers and tunnelling, the first-aiders should be trained to cope with foreseeable emergencies that may be beyond the scope of the standard training course. Where there is shift working the contractor should ensure that sufficient first-aiders are appointed to provide adequate coverage for each shift.

First-aid training

The training of first-aiders and the subsequent qualifications given by the British Red Cross Society, St John Ambulance Association and St Andrews Ambulance Association are approved by the Health and Safety Executive. Details of the courses and the centres where they are held may be obtained from the British Red Cross Society in London, the St John Ambulance in London and the St Andrews Ambulance Association in Glasgow.

The statutory certificate course for first-aiders takes four consecutive days and covers

- resuscitation
- control of bleeding
- treatment of shock
- recognition of illness
- treatment of injuries
- simple record-keeping
- dealing with poisoning
- treatment of the unconscious patient
- dressing and immobilisation of injured parts
- contents of first-aid boxes and their uses
- transport of sick and injured patients
- treatment of burns and scalds
- personal hygiene when dealing with wounds
- communications and delegation in an emergency.

The statutory first-aid certificate lasts for three years, after which a requalification course is necessary. However, one-day refresher courses are available for those who wish to update their training every year. This is also recommended for appointed persons.

One-day courses are available for appointed persons at a number of centres. The courses give training in the handling of emergencies and the priorities for the care of a casualty when first aid is being rendered. The syllabus covers

- priorities for first aid
- asphyxia
- resuscitation
- shock and blood loss
- burns and scalds
- treatment of the unconscious casualty
- treatment of heart attacks
- treatment of strokes
- common workplace injuries.

On-site training of first-aiders is illustrated in Fig. 11.3.

Fig. 11.3.   St John Ambulance trainers give first aid to 'casualties' on a construction site

Selection of staff for first-aid training

In terms of the turnover of most companies involved in the construction industry, the cost of first-aid training is relatively small and it is expected that most firms will send many more employees for first-aid training than the statutory minimum. It is customary for first-aiders to receive a small salary increment in recognition of their special duties and occasional personal time involved in reading technical literature and other activities relating to first aid. It is recommended that chartered engineers and builders, and especially those on construction sites, should attend a first-aiders' course or at least the appointed person's course. Apart from providing relief of suffering to the injured, the knowledge helps those in managerial positions to stay calm and act in a positive and sympathetic manner when an accident occurs.

First aid

First aid is a post-accident strategy vital to prevent loss of life and future deterioration in health following accidental injury or sudden illness. First aid is defined as the skilled application of accepted principles of treatment on the occurrence of an accident or in the case of sudden illness

- to sustain life
- to prevent deterioration in an existing condition
- to promote recovery.

The most important areas of first-aid treatment are

- restoration of breathing (resuscitation)
- control of bleeding
- prevention of collapse.

First-aiders should take care not to become casualties themselves, for example, by entering a dangerous atmosphere without breathing apparatus or by entering a collapsed excavation without protection to prevent further collapse. Protective clothing should be worn and safety equipment used to reduce the risk of further injuries. Where help or an ambulance is needed it should be sent for without delay.

Simple first-aid procedures

As many employees as possible should be trained in simple first-aid procedures, namely resuscitation, control of bleeding and treatment of the unconscious patient, so as to ensure that casualties receive prompt attention. A number of simple first-aid procedures are outlined in the following paragraphs (acknowledgements to *RoSPA health and safety practice* by Stranks and Dewis[2]).

Resuscitation. Resuscitation (artifical respiration) may be required where the victim has suffered electric shock, gassing, suffocation or drowning. In each case, certain preliminary action is necessary as follows.

- *Electric shock.* Switch off the current if possible, otherwise pull the victim from contact, using heavy-duty insulating gloves, a rubber sheet, piece of dried timber, cloth, a folded newspaper, rope, the victim's own clothing if dry or other dry non-conducting material. Extreme care must be taken not to touch the victim's skin before the current is switched off.
- *Gassing.* Remove.the victim to the fresh air or a gas-free atmosphere as quickly as possible, ensuring that the rescuer is wearing suitable respiratory protection (e.g. breathing apparatus).
- *Suffocation.* If the victim has been buried in loose materials (e.g. in a trench collapse), immediately and quickly clear any debris from his mouth and nose. It may be desirable in this case to begin mouth-to-mouth resuscitation immediately the victim's head has been uncovered.
- *Drowning.* Remove the victim from the water with all speed. Clear any debris from his mouth. If rescue is by boat begin resuscitation in the boat. When the victim is clear of danger, resuscitation should be started immediately as the first minutes are vital, and should be continued without interruption until breathing is restored or until a doctor certifies that life is extinct. Resuscitation procedure is described in Fig. 11.4.

Fig. 11.4. Resuscitation procedure: (a) place the casualty on his back and clear his mouth; (b) tilt his head to open airway while supporting his jaw; (c) kneel beside casualty, open his mouth and pinch his nose; (d) open your mouth, take a deep breath, seal his mouth with yours and breathe firmly into it; when you see casualty's chest rise, remove your mouth and let his chest fall; continue until the casualty is breathing

Bleeding. Where bleeding is more than minimal, control it by direct pressure. Apply a pad of sterilised dressing or if necessary direct pressure with the fingers or thumb on the bleeding point. Raising a limb if the bleeding is sited there will help to reduce the flow of blood (unless the limb is fractured).

Unconsciousness. Where the patient is unconscious, care must be taken to keep the airways open. This is done by clearing the mouth and ensuring that the tongue does not block the back of the throat. Where possible the patient should be placed in the recovery position (*see* Fig. 11.5).

Broken bones. Unless in a position which exposes the victim to further danger, no attempt should be made to move a casualty with suspected broken bones or injured joints until the injured parts have been supported so that they cannot move or be moved separately (from the rest of the body).

Burns and scalds. Burns and scalds should be treated by flushing the affected area with plenty of clean cool water before applying a sterilised dressing or a clean towel. Where the burn is large or deep, a sterilised dressing should be applied. It is most important that blisters do not burst and that no attempt is made to remove clothing which may be sticking to burns or scalds. In the case of chemical burns any contaminated clothing which shows no signs of sticking to the skin should be removed, and all affected parts of the body flushed with plenty of cool clean water, ensuring that all the chemical substance is so diluted as to be rendered harmless. A sterilised dressing should be applied to exposed damaged skin and clean towels to damaged areas where the clothing cannot be removed. Care should be taken when treating such a casualty to avoid personal contamination.

Foreign bodies in the eye. Unless the object can be removed easily with a clean piece of moist material, no attempt should be made to remove the object. The eye should be irrigated with clean cool water. People with eye injuries or foreign bodies in the eye which cannot be removed in simple fashion should always be sent to hospital immediately, with the eye covered by a sterilised eye-pad dressing. It is now well-established medical opinion that no first aiders should attempt to remove any form of foreign body from the eye of an injured person.

Chemical substances in the eye. The open eye should be flushed at once with cool clean water, the treatment continuing for five to ten minutes or even longer. If the contamination is more than minimal the casualty should be taken to hospital.

AIDS. Although the risk of first-aiders contracting AIDS or other communicable diseases while treating patients is extremely small, recommendations are published by the Health and Safety Executive in the Construction Industry Advisory Committee health hazard information sheet No. 6 *AIDS* (*see* Appendix 4). First-aiders should cover their own cuts or abrasions with waterproof dressings before treating a patient. Wounds should be thoroughly irrigated and washed with soap and water before they are dressed. First-aiders should always wash their hands after treating a wound. Although no cases of AIDS have been reported anywhere by transmission from saliva, a portable mouth-to-mouth resuscitation device incorporating a one-way valve can be used.

Fig. 11.5. Put the casualty in the recovery position

Reporting injuries

Reporting an injury or dangerous occurrence

The Reporting of Injuries, Diseases and Dangerous Occurrences Regulations 1995, often referred to as RIDDOR 95, require the notification in writing to the enforcing authority of all injuries resulting from accidents at work which cause incapacity for more than three days. In addition, any of the following events must first be notified to the enforcing authority by the quickest practicable means

- the *death* of any person as a result of an accident in connection with work
- any person suffering any of the following injuries or conditions as a result of an accident in connection with work; these are otherwise referred to as *reportable major injuries*
  - fractures other than to fingers, thumbs or toes
  - amputation
  - dislocation of the shoulder, hip, knee or spine
  - loss of sight (temporary or permanent)
  - chemical or hot metal burn to the eye or any penetrating injury to the eye
  - injury resulting from an electric shock or electrical burn leading to unconsciousness or requiring resuscitation or admittance to hospital for more than 24 hours
  - any other injury: leading to hypothermia, heat-induced illness or unconsciousness; or requiring resuscitation; or requiring admittance to hospital for more than 24 hours
  - unconsciousness caused by asphyxia or exposure to harmful substance or biological agent
  - acute illness requiring medical treatment, or loss of consciousness arising from absorption of any substance by inhalation, ingestion or through the skin
  - acute illness requiring medical treatment where there is reason to believe that this resulted from exposure to a biological agent or its toxins or the infected material.
- any of the *dangerous occurrences* listed in Appendix 5.

In the case of a death, a specified major injury or dangerous occurrence notification must be made by telephone. This must be followed up within seven days by a written report to the enforcing authority on form F2508, available from HMSO. The enforcing authority is the body responsible for enforcing the HSW Act on the premises where or in connection with the work at which the reportable injury or dangerous occurrence happened. It may thus be either the Health and Safety Executive or the local authority. Construction site injuries and dangerous occurrences should be reported to the area office of the Health and Safety Executive applicable to the site address (*see* Appendix 3).

CDM Regulations

On contracts where the CDM Regulations apply contractors making reports under RIDDOR must also immediately inform the principal contractor who may wish to investigate the incident and consider whether any changes to the health and safety plan are necessary (*see* p. 25).

Keeping records

A record must be made and kept of all reportable injuries and dangerous occurrences. The record must contain in each case the following information

- the date and time of the accident causing the injury or dangerous occurrence
- the following particulars about the person affected
  - full name
  - occupation
  - nature of injury or condition
- the place where the accident or dangerous occurrence happened
- a brief description of the circumstances.

Reporting a disease

Diseases are reported on form F 2508A. *Everyone's guide to RIDDOR '95* (HSE 31), available free from the Health and Safety Executive (*see* Appendix 4), provides a brief guide to the reporting and record-keeping of cases of disease.

Under the regulations, certain listed diseases must be reported either by employers or by the self-employed when those diseases are linked to specified types of work. If an employer is notified by a doctor that one of his employees suffers from a reportable work-related disease then a report form F2508A must be completed and sent to the enforcing authority. Reportable diseases include

- certain poisonings
- some skin diseases such as occupational dermatitis, skin cancer, chrome ulcer, oil folliculitis/acne
- lung diseases including occupational asthma, farmer's lung, pneumoconiosis, asbestosis, mesothelioma
- infections such as leptospirosis, hepatitis, tuberculosis, anthrax, legionellosis and tetanus
- other conditions such as occupational cancer, certain muscuoskeletal disorders, decompression illness and hand–arm vibration syndrome.

Diseases related to construction work should always be reported to the Health and Safety Executive.

Internal organisation for reporting

It will be necessary for many of the larger companies in the construction industry to issue instructions for the reporting of injuries and dangerous occurrences to the enforcing authority. The company itself will need to define which person completes form F2508 or F2508A and whose responsibility it is to initiate the investigation, inform relatives and friends and deal with other administrative matters.

Accident investigation

The causes of accidents should be investigated

- to identify the cause and then to initiate measures to prevent a recurrence
- to gather information for use in any criminal or civil proceedings
- to confirm or refute a claim for industrial injury benefit
- to prepare notifications to be made to the enforcing authority.

It is vital that all accidents are investigated as quickly as possible. The longer the delay the less likely it is that the true facts will be ascertained. It is seldom that the cause of an accident is simple and it is equally seldom that only one or two people are to blame. Usually an accident arises from a number of acts or omissions by a number of people and the investigation should therefore be conducted in a calm, orderly atmosphere that encourages the witnesses to talk freely.

Investigation procedure

It is often necessary to conduct two or more interviews with witnesses. The first investigation will attempt to determine

- what caused the accident
- who was involved
- when, where and why it occurred
- how it could have been prevented.

It may then be necessary to consider obtaining further information to establish whether employees were adequately trained and whether proper work systems were being followed and, if not, how they could be improved to prevent a recurrence.

The cause of an accident should never be classified as carelessness because the only remedy to a careless act is for more care to be taken in the future. It is essential to be specific and to define the act or omission that caused the accident.

References

Chapter 1

1. HEALTH AND SAFETY EXECUTIVE. *Blackspot construction* HSE, London, 1988.
2. HEALTH AND SAFETY COMMISSION. *Designing for health and safety in construction. A guide for designers on the Construction (Design and Management) Regulations 1994.* HSE, London, 1995.

Chapter 2

1. BISHOP D. *Opportunities and impositions – an outline of the CDM Regulations and their impact upon the role of construction professionals.* Construction Industry Council, London, 1995.
2. JOYCE R. *The CDM Regulations explained.* Thomas Telford, London, 1995.
3. JOYCE R. *The C(H,S&W) Regulations explained.* Thomas Telford, London, 1997.

Chapter 4

1. TOMLINSON M.J. General safety in excavation. *Hazards in construction.* Institution of Civil Engineers, London, 1971, 31–39.
2. TIMBER RESEARCH AND DEVELOPMENT ASSOCIATION. *Timber in excavation.* TRADA, London, 1990.
3. BRITISH STANDARDS INSTITUTION. *Code of practice for access and working scaffolds and special scaffold structures in steel.* BSI, London, 1993, BS 5973.
4. BRITISH STANDARDS INSTITUTION. *Code of practice for falsework.* BSI, London, 1982, BS 5975.
5. BRITISH STANDARDS INSTITUTION. *Code of practice for the use of safety nets, containment nets and sheets on construction works.* BSI, London, 1991, BS 8093.
6. BRITISH STANDARDS INSTITUTION. *Code of practice for the safe use of cranes (mobile cranes, tower cranes and derrick cranes).* BSI, London, 1972, CP 3010.
7. HEALTH AND SAFETY EXECUTIVE. *Blackspot construction.* HSE, London, 1988.
8. BRITISH STANDARDS INSTITUTION. *Rollover protective structures on earth moving machinery. Specification for crawler wheel loaders and tractors, backhoe loaders, graders, tractor scrapers, articulated shear dumpers.* BSI, London, 1987, BS 5527: Part 1.
9. BRITISH STANDARDS INSTITUTION. *Specification for falling object protective structures on earth moving machinery: Laboratory tests and performance requirements.* BSI, London, 1992, BS 6912: Part 6.
10. COUNTY SURVEYOR'S SOCIETY – DEPARTMENT OF TRANSPORT. *Notes for guidance on safety at roadworks.* County Surveyor's Society, Wiltshire, 1994.
11. DEPARTMENT OF TRANSPORT, SCOTTISH DEVELOPMENT OFFICE AND WELSH OFFICE. *Traffic signs manual.* HMSO, London, 1974.

12. DEPARTMENT OF TRANSPORT, SCOTTISH OFFICE AND WELSH OFFICE. *Safety at street works and road works. A code of practice issued under sections 65 and 124 of the New Roads and Street Works Act 1991.* HMSO, London, 1993.
13. BRITISH STANDARDS INSTITUTION. *Code of practice for safety in tunnelling in the construction industry.* BSI, London, 1990, BS 6164.
14. BRITISH STANDARDS INSTITUTION. *Code of practice for demolition.* BSI, London, 1982, BS 6187.
15. BRITISH STANDARDS INSTITUTION. *Code of practice for the safe use of explosives in the construction industry.* BSI, London, 1988, BS 5607.
16. BRITISH STANDARDS INSTITUTION. *Life jackets and personal buoyancy aids additional items.* BSI, London, 1994, BS EN 394 and *Life jackets and personal buoyancy aids. Life jacket 150.* BSI, London, 1994, BS EN 396.
17. HEALTH AND SAFETY EXECUTIVE. *Noise at work.* HMSO, London, 1989.
18. HEALTH AND SAFETY EXECUTIVE. *The protection of persons against ionising radiations arising from any work activity.* HSE, London, 1994.

Chapter 6

1. WATER RESEARCH CENTRE. *Sewers for adoption: a design and construction guide for developers.* WRC, Henley, 1985.
2. HEALTH AND SAFETY EXECUTIVE. *Blackspot construction.* HSE, London, 1988.
3. BRITISH STANDARDS INSTITUTION. *Code of practice for demolition.* BSI, London, 1982, BS 6187.
4. CONSTRUCTION INDUSTRY RESEARCH AND INFORMATION ASSOCIATION. *Quality assurance in civil engineering.* CIRIA, London, 1985, CIRIA report 109.

Chapter 7

1. ABBOTT H. *Safer by design.* Design Council, London, 1987.

Chapter 10

1. HEALTH AND SAFETY EXECUTIVE. *Noise at work.* HMSO, London, 1989.
2. BRITISH STANDARDS INSTITUTION. *Specification for eye protectors for industrial and non-industrial use.* BSI, London, 1987, BS 2092.
3. BRITISH STANDARDS INSTITUTION. *Specification for equipment for eye, face and neck protection against non-ionising radiation arising during welding and similar operations.* BSI, London, 1982, BS 1542.
4. BRITISH STANDARDS INSTITUTION. *Specification for filters, cover lenses and backing lenses for use during welding and similar operations.* BSI, London, 1989, BS 679.
5. BRITISH STANDARDS INSTITUTION. *Specification for filtering half masks to protect against particles.* BSI, London, 1992, BS EN 149.
6. BRITISH STANDARDS INSTITUTION. *Specification for high visibility warning clothing.* BSI, London, 1994, BS EN 471.
7. BRITISH STANDARDS INSTITUTION. *Specification for industrial safety belts, harnesses and safety lanyards.* BSI, London, BS 1397.

Chapter 11

1. HEALTH AND SAFETY EXECUTIVE. *First aid at work. Health and Safety (First Aid) Regulations 1981.* HSE Publications, London, 1990, Approved code of practice.
2. STRANKS J. AND DEWIS M. *RoSPA health and safety practice.* Pitman, London, 1986.

Appendix 1
A selection of acts and statutory instruments relevant to the construction industry

Amendments and the dates of amendments are not given.

Mines and Quarries Act 1954
Factories Act 1961
Work in Compressed Air Special Regulations 1958 and 1960[1]
Construction (General Provisions) Regulations 1961[2]
Construction (Lifting Operations) Regulations 1966[3]
Abrasive Wheels Regulations 1970
Highly Flammable Liquids and Liquefied Petroleum Gases Regulations 1972
Woodworking Machines Regulations 1974
Offices, Shops and Railway Premises Act 1963
Control of Pollution Act 1974
Health and Safety at Work etc. Act 1974
Safety Signs Regulations 1980
Control of Lead at Work Regulations 1980
Health and Safety (First aid) Regulations 1981
Diving Operations at Work Regulations 1981
Ionising Radiations Regulations 1985
Control of Asbestos at Work Regulations 1987
Quarries (Explosives) Regulations 1988
Control of Substances Hazardous to Health Regulations 1988
Construction (Head Protection) Regulations 1989
Electricity at Work Regulations 1989
Noise at Work Regulations 1989
Control of Explosives Regulations 1991
Management of Health & Safety at Work Regulations 1992
Manual Handling Operations Regulations 1992
Personal Protective Equipment Regulations 1992
Provision and Use of Work Equipment Regulations 1992
Reporting of Injuries Diseases and Dangerous Occurrences Regulations 1995
Construction (Health, Safety and Welfare) Regulations 1996
Confined Spaces Regulations (in preparation, 1996)
Workplace (Fire Precautions) Regulations (in preparation, 1996)

Lifting Operations Regulations (in preparation, 1996)[3]
Highways Act 1980
Building Act 1984
Building Regulations 1985
New Roads and Street Works Act 1991

Notes

1. Work in Compressed Air Special Regulations are due for amendment in 1996.
2. Only Regulations 42, 43 and 57 of the Construction (General Provisions) Regulations 1961 remain in force in 1996. The Regulations will be finally revoked in 1997
3. The Construction (Lifting Operations) Regulations 1966 are to be updated and consolidated within new Lifting Operations Regulations (in preparation, 1996).

Appendix 2
Building Regulations 1985 made under the Building Act 1984

Approved document	Subject covered
A	Structure Loading and ground movement Disproportionate collapse
B	Fire safety: means of escape fire spread, access and facilities for the fire service
C	Preparation of site: dangerous and offensive substances, subsoil drainage resistance to weather and ground moisture
D	Toxic substances: cavity insulation
E	Sound: airborne and impact sound
F	Ventilation
G	Hygiene: sanitary conveniences and washing facilities, bathrooms, hot water storage
H	Drainage and waste disposal: sanitary pipework and drainage, cesspools and tanks rainwater drainage, solid waste drainage
J	Heat producing appliances
K	Stairs, ramps and guards, vehicle barriers
L	Conservation of fuel and power
M	Access and facilities for disabled people
N	Glazing: materials and protection

Appendix 3
Health and Safety Executive addresses

Headquarters
Health and Safety Commission, Rose Court, 2 Southwark Bridge London SE1 9HS
(Tel 0171 717 6000)

Area offices

Area	Address	Tel/Fax	Local authorities covered
South West	InterCity House Mitchell Lane Victoria Street Bristol BS1 6AN	Tel: 0117 9290681 Fax: 0117 9262998	Avon, Cornwall, Devon, Gloucestershire, Somerset, Isles of Scilly
South	Priestley House Priestley Road Basingstoke RG24 9NW	Tel: 01256 404000 Fax: 01256 404100	Berkshire, Dorset, Hampshire, Isle of Wight, Wiltshire
South East	3 East Grinstead House, London Road East Grinstead West Sussex RH19 1RR	Tel: 01342 326922 Fax: 01342 312917	Kent, Surrey, East Sussex, West Sussex
London North	Maritime House 1 Linton Road Barking Essex IG11 8HF	Tel: 0181 594 5522 Fax: 0181 591 5183	Barking and Dagenham Barnet, Brent, Camden, Ealing, Enfield, Hackney, Haringey Harrow, Havering Islington, Newham, Redbridge, Tower Hamlets, Waltham Forest

Area	Address	Tel/Fax	Local authorities covered
London South	1 Long Lane London SE1 4PG	Tel: 0171 407 8911 Fax: 0171 403 7058	Bexley, Bromley, City of London, Croydon, Greenwich, Hammersmith & Fulham, Hillingdon, Hounslow, Kensington & Chelsea, Kingston, Merton, Richmond, Southwark, Sutton, Wandsworth, Westminster
East Anglia	39 Baddow Road Chelmsford CM2 0HL	Tel: 01245 284661 Fax: 01245 252633	Essex, Norfolk, Suffolk
Northern Home Counties	14 Cardiff Road Luton, Beds. LU1 1PP	Tel: 01582 444200 Fax: 01582 459775	Bedfordshire, Buckinghamshire, Cambridgeshire, Hertfordshire
East Midlands	5th Floor, Belgrave House, 1 Greyfriars, Northampton NN1 2BS	Tel: 01604 738300 Fax: 01604	Leicestershire, Northamptonshire, Oxfordshire, Warwickshire
West Midlands	McLaren Building 2 Masshouse Circus Queensway Birmingham B4 7NP	Tel: 0121 609 5200 Fax: 0121 609 5349	Birmingham, Coventry Dudley, Sandwell Solihull, Walsall, Wolverhampton
Wales	Brunel House 2 Fitzalan Road Cardiff CF2 1SH	Tel: 01222 473777 Fax: 01222 473642	Clwyd, Dyfed, Gwent, Gwynedd, Mid Glamorgan, Powys, South Glamorgan, West Glamorgan
Marches	The Marches House Midway, Newcastle under Lyme, Staffs ST5 1DT	Tel: 01782 717181 Fax: 01782 620612	Hereford and Worcester, Shropshire, Staffordshire
North Midlands	Birbeck House Trinity Square Nottingham NG1 4AU	Tel: 01159 470710 Fax: 01159 411577	Derbyshire, Lincolnshire, Nottinghamshire

Area	Address	Tel/Fax	Local authorities covered
South Yorkshire	Sovereign House 110 Queen Street SHEFFIELD S1 2ES	Tel: 01142 739081 Fax: 01142 755746	Barnsley, Doncaster Humberside Rotherham Sheffield
West and North Yorkshire	8 St Pauls Street Leeds LS1 2LE	Tel: 01132 446191 Fax: 01132 450626	Bradford, Calderdale Kirklees, Leeds Wakefield Craven Hambleton, Harrogate Richmondshire, Ryedale Scarborough, Selby York
Greater Manchester	Quay House Quay Street Manchester M3 3JB	Tel: 0161 831 7111 Fax: 0161 831 7169	Bolton, Bury, Manchester Salford, Oldham, Rochdale, Stockport, Tameside, Trafford Wigan
Merseyside	The Triad Stanley Road Bootle Merseyside	Tel: 0151 922 7211 Fax: 0151 922 5031	Cheshire, Merseyside
North West	Victoria House Ormskirk Road Preston PR1 1HH	Tel: 01772 259321 Fax: 01772 821807	Cumbria, Lancashire
North East	Arden House Regent Centre Regent Farm Road Gosforth Newcastle upon Tyne NE3 3JN	Tel: 0191 284 8448 Fax: 0191 285 9682	Cleveland, Durham, Northumberland, Tyne and Wear
Scotland East	Belford House 59 Belford Road Edinburgh EH4 3UE	Tel: 0131 247 2000 Fax: 0131 247 2121	Borders, Central, Fife, Grampian, Highland, Lothian, Tayside and the island areas of Orkney and Shetland
Scotland	375 West George Street GLASGOW G2 4LW	Tel: 0141 275 3000 Fax: 0141 275 3100	Dumfries and Galloway, Strathclyde and the Western Isles

Health and Safety Executive Information Centres

Sheffield Information Centre
Broad Lane
Sheffield S3 7HQ
Tel 01142 892345
Fax 01142 892333

HSE Infoline
Tel: 0541 545 500

Bootle Information Centre
St Hugh's House
Stanley Precinct
Bootle, Merseyside L20 3QY
(personal callers only)

London Information Centre
Rose Court
2 Southwark Bridge
London SE1 9HS
(personal callers only)

Source: Health and Safety Executive 1995

Appendix 4
A selection of HSE books relevant to construction

Abbreviations

CRR	Contract research reports
CS	Chemical safety
EH	Environmental hygiene
GS	General series
PM	Plant and machinery
HS(G)	Health and safety guidance
COP	Approved codes of practice
HS(R)	Guidance on health and safety regulations
L	Legal

Chapter 1

Accident research

Blackspot construction: a study of five years fatal accidents in the building and civil engineering industries. 1988 ISBN 0 11 883992 6

Deadly maintenance: study of fatal accidents at work. 1985 ISBN 0 11 883806 7

HS(G)96
The costs of accidents at work. 1993 ISBN 0 7176 0439 X

GS 7 (Rev)
Accidents to children on construction sites. 1989 ISBN 0 11 885416 X

CRR 45/1992
Research into management, organisational and human factors in the construction industry. 1992 ISBN 0 11 886383 5

CRR 51/1993
Improving safety on construction sites by changing personnel behaviour. 1993 ISBN 0 11 8821482

Chapter 2

L1
A guide to the Health and Safety at Work etc Act 1974. 4th ed 1990 ISBN 0 7176 0441 1

L54
Managing construction for health and safety. Construction (Design and Management) Regulations 1994. Approved code of Practice. ISBN 0 7176 0755 0

HS(G)150
Health and safety in construction. Guidance for everyone in construction work. ISBN 0 7176 1143 4

IND(G)220L
CHSW Regulations. A guide to the Construction (Health, Safety and Welfare) Regulations 1996. ISBN 0 7176 1161 2

Protecting the public – your next move. ISBN 0 7176 1148 5

Chapter 4

General site operations

HS(G)46
A guide for small contractors. Site safety and concrete construction. 1989 ISBN 0 11 885475 5

HS(G)130
Health and safety for small construction sites. 1995 ISBN 0 7176 0806 9

HS(G)150
Health and safety in construction. Guidance for everyone in construction work. ISBN 0 7176 1143 4

GS 49
Pre-stressed concrete. 1991 ISBN 0 11 885597 2

PM 46
Wedge and socket anchorages for wire ropes. 1985 ISBN 0 11 883611 0

HS(G)17
Safety in the use of abrasive wheels. 1992 ISBN 0 7176 0466 7

HS(G)5
Hot work: welding and cutting on plant containing flammable materials. 1979 ISBN 0 11 883229 8

GS 4(Rev)
Safety in pressure testing. 1992 ISBN 0 11 886338 X

PM 17
Pneumatic nailing and stapling tools. 1979 ISBN 0 11 883192 5

L4
Woodworking Machines Regulations 1974. Guidance on regulations. 1991 ISBN 0 11 885592 1

L22
Work equipment. Provision and Use of Work Equipment Regulations 1992. Guidance on regulations. 1992 ISBN 0 7176 0414 4

Electricity on site

HS(G)85
Electricity at work: safe working practices. 1993 ISBN 0 7176 0442 X

HS(g)141
Electrical safety on construction sites. 1995 ISBN 0 7176 1000 4

PM 32
Safe use of portable electrical apparatus. 1990 ISBN 0 11 885590 5

HS(G)107
Maintaining portable and transportable electrical equipment. 1994 ISBN 0 7176 0715 1

HS(R)25
Memorandum of guidance on the Electricity at Work Regulations 1989. 1989 ISBN 0 11 883963 2

Lighting

HS(G)38
Lighting at work. 1987 ISBN 0 7176 0467 5

Dangerous substances

CS 4
Keeping of LPG in cylinders and similar containers. 1986 ISBN 0 7176 0631 7

CS 6
Storage and use of LPG on construction sites. 1981 ISBN 0 11 883391 X

HS(G)34
Storage of LPG at fixed installations. 1987 ISBN 0 11 883908 X

L38
Approved guide to the classification and labelling of substances and preparations dangerous for supply. 1993 ISBN 0 11 882155 5

HS(G)50
The storage of flammable liquids in fixed tank (up to 10 000 m^3 total capacity). 1990 ISBN 0 11 885532 8

L53
Approved methods for the classification and packaging of dangerous goods for carriage by road and rail. 1994 ISBN 0 7176 0744 5

L56
The Gas Safety (Installations and Use) Regulations. 1994 Approved code of practice. ISBN 0 7176 0797 6

COP 27
Explosives at quarries. Quarries (Explosives) Regulations 1988. Approved code of practice. 1989 ISBN 0 11 885462 3

HS(G)113 (new entry)
Lift trucks in hazardous areas. 1994 ISBN 0 7176 0706 2

HS(G)37
Introduction to local exhaust ventilation. 1993 ISBN 0 11 882134 2

Scaffolding

GS 15
General access scaffolds. 1982 ISBN 0 11 883545 9

GS 31
Safe use of ladders, step ladders and trestles. 1984 ISBN 0 11 883594 7

GS 42
Tower scaffolds. 1987 ISBN 0 11 883941 1

PM 30
Suspended access equipment. 1983 ISBN 0 11 883577 7

Falsework
HS(G)32
Safety in falsework for in situ beams and slabs. 1987 ISBN 0 11 883900 4

Structures
GS 28/1
Safe erection of structures. Part 1: initial planning and design. 1984 ISBN 0 11 883584 X

GS 28/2
Safe erection of structures. Part 2: site management and procedures. 1985 ISBN 0 11 883605 6

GS 28/3
Safe erection of structures. Part 3: working places and access. 1986 ISBN 0 11 883530 0

GS 28/4
Safe erection of structures. Part 4: legislation and training. 1986 ISBN 0 11 883531 9

Lifting operations and plant
GS 6 (Rev)
Avoidance of danger from overhead electrical lines. 1991 ISBN 0 11 885668 5

HS(G)47
Avoiding danger from underground services. 1989 ISBN 0 7176 0435 7

HS(G)23
Safety at power operated mast work platforms. 1985 ISBN 0 11 883820 2

PM 42
Excavators used as cranes. 1984 ISBN 0 11883591 2

PM 54
Lifting gear standards. 1985 ISBN 0 11 883520 3

PM 55
Safe working with overhead travelling cranes. 1985 ISBN 0 11 883524 6

PM 63
Inclined hoists used in building and construction work. 1987 ISBN 0 11 883945 4

HS(G)6
Safety in working with lift trucks. 1993 ISBN 0 11 886395 9

L20
A guide to the Lifting Plant and Equipment (Records of Test and Examination etc) Regulations 1992. 1992 ISBN 0 7176 0488 8

HS(G)136
Workplace transport safety. Guidance for employers. 1995 ISBN 0 7176 0935 9

Sewers and confined spaces

GS5
Entry into confined spaces. 1995 ISBN 0 7176 0787

Demolition

GS 29/1 (Rev)
Health and safety in demolition work. Part 1: preparation and planning. 1988 ISBN 0 11 885405 4

GS 29/3
Health and safety in demolition work. Part 3: techniques. 1984 ISBN 0 11 883609 9

GS 29/4
Health and safety in demolition work. Part 4: health hazards. 1985 ISBN 0 11 883604 8

Asbestos and dust

EH 40/94
Occupational exposure limits. 1994 ISBN 0 7176 0722 4

EH40/94
Workplace air and biological monitoring database: instrument enhanced. This has been developed by HSE using d-Base III and is available on floppy disc in both 5¼ and 3½ inch format. 2 vols and envelope with 2 computer diskettes. ISBN 0 7176 0736 4

EH 64
Occupational exposure limits: criteria document summaries. Synopses of the data used in setting occupational exposure limits. 1994 ISBN 0 7176 0759 3
NB: summaries should be read in conjunction with the COSHH Regulations 1988, associated Approved Codes of Practice, and guidance publications EH 40 and EH 42

L27
The Control of Asbestos at Work. Control of Asbestos at Work Regulations 1987. Approved code of practice. 2nd ed 1993 ISBN 0 11 882037 0

EH 35 (Rev)
Probable asbestos dust concentrations at construction processes. 1989 ISBN 0 11 885421 6

EH 36 (Rev)
Work with asbestos cement. 1990 ISBN 0 11 885422 4

EH 37 (Rev)
Work with asbestos insulating board. 1989 ISBN 0 11 885423 2

EH 10 (Rev)
Asbestos – exposure limits and measurement of airborne dust concentrations. 1990 ISBN 0 11 885552 2

L11
A guide to the Asbestos (Licensing) Regulations 1983. 1991 ISBN 0 11 885684 7

EH 52
Removal techniques and associated waste handling for asbestos insulation, coatings and insulating board. 1989 ISBN 0 11 885409 7

EH 47 (Rev)
Provision, use and maintenance of hygiene facilities for work with asbestos insulation and coatings. 1990 ISBN 0 11 885567 0

HS(G)103 (new entry)
Safe handling of combustible dusts. 1994 ISBN 0 7176 0725 9

EH 65/12
Portland cement dust. 1994 ISBN 0 7176 0763 1

EH 65/2
Pulverised fuel ash: criteria document for an occupational exposure limit. 1992 ISBN 0 11 886391 6

Explosives
L10
A guide to the Control of Explosives Regulations 1991. 1991 ISBN 0 11 885670 7

HS(G)114 (new entry)
Conditions for the authorisation of explosives in Great Britain. 1994 ISBN 0 7176 0717 8

HS(G)36
Disposal of explosives waste and the decontamination of explosives plant. 1987 ISBN 0 11 883926 8

Diving and compressed air
L6
Diving operations at work. The Diving Operations at Work Regulations 1981 as amended by the Diving Operations at Work (Amendment) Regulations 1990. 1991 ISBN 0 11 885599 9

HS(G)39 (Rev)
Compressed air safety. 1990 ISBN 0 11 885582 4

Health hazards

Health
L5
Control of substances hazardous to health and control of carcinogenic substances. Control of Substances Hazardous to Health Regulations 1994. Approved codes of practice. 1995 ISBN 0 7176 0819 0

HS(G)97
A step by step guide to COSHH assessment. 1993 ISBN 0 11 886379 7

HS(G)137
Health risk management: a practical guide for managers in small and medium sized enterprises. 1995 ISBN 0 7176 0905 7

HS(G)77
COSHH and peripatetic workers. 1992 ISBN 0 11 885733 9

HS(G)61
Surveillance of people exposed to health risks at work. 1990 ISBN 0 11 885574 3

L55
Preventing asthma at work. 1994 ISBN 0 7176 0661 9

HS(G)110 (new entry)
Seven steps to successful substitution of hazardous substances. 1994 ISBN 0 7176 0695 3

Asbestos
see Demolition, p. 290

Ionising Radiations

L58
The protection of persons against ionising radiation arising from any work activity. 1994 ISBN 0 7176 0508 6

COP 16 (*see also* L7)
Protection of persons against ionising radiation arising from any work activity: The Ionising Radiations Regulations 1985: approved code of practice. 1985 ISBN 0 7176 0508 6

HS(G)91
A framework for the restriction of occupational exposure to ionising radiation. 1992 ISBN 0 11 886324 X

L7 (*see also* COP 23)
Dose limitation – restriction of exposure: Additional guidance on regulation 6 of the Ionising Radiations Regulations 1985. Approved code of practice – part 4. 1991 ISBN 0 11 885605 7

Use of wood preservatives

Remedial timber treatment in buildings: a guide to good practice and the safe use of wood preservatives. 1991 ISBN 0 11 885987 0

Lead at work

COP 2
Control of lead at work: approved code of practice: revised June 1985 (in support of SI 1980 No 1248) ISBN 0 11 883780 X

Noise

Noise at work. Noise guide No 1: Legal duties of employers to prevent damage to hearing. Noise guide No 2: legal duties of designers, manufacturers, importers and suppliers to prevent damage to hearing. The Noise at Work Regulations 1989. 1989 ISBN 0 11 885512 3

HS(G)56
Noise at work. Noise assessment, information and control. Noise guides 3 to 8. 1990 ISBN 0 11 885430 5

HS(G)109
Control of noise in quarries. 1993 ISBN 0 7176 0648 1

CRR 68/1995
Occupational hearing loss from low level noise. 1995 ISBN 0 7176 0885 9

VDUs

L26
Display screen equipment work. Health and Safety (Display Screen Equipment) Regulations 1992. Guidance on regulations. 1992 ISBN 0 7176 0410 1

HS(G)90
VDUs: an easy guide to the Regulations. 1994 ISBN 0 7176 0735 6

Ergonomic matters

L23
Manual handling. Manual Handling Operations Regulations 1992. Guidance on regulations. 1992 ISBN 0 7176 0411 X

HS(G)115
Manual handling – solutions you can handle. 1994 ISBN 0 7176 06937

HS(G)121
A pain in your workplace: ergonomic problems and solutions. 1994 ISBN 0 7176 0668 6

HS(G)60
Work related upper limb disorders: a guide to prevention. 1990 ISBN 0 7176 0475 6

HS(G)88
Hand-arm vibration. 1994 ISBN 0 7176 0743 7

Chapter 6

Management

L1
A guide to the Health and Safety at Work etc Act 1974. 4th ed 1990 ISBN 0 7176 0441 1

L21
Management of Health and Safety at Work Regulations 1992. Approved code of practice. 1992 ISBN 0 7176 0412 8

COP 1
Safety representatives and safety committees. 1988 (The Brown Book) ISBN 0 11 883959 4

HS(G)65
Successful health and safety management. 1991 ISBN 0 7176 0425 X

U10
Designing for health and safety in construction. 1995 ISBN 0 7176 0807 7

CRR 72/1995
Information on site safety for designers of smaller building projects. 1995 ISBN 0 7176 0777 1

L54
Managing construction for health and safety Construction (Design and Management) Regulations 1994. Approved Code of Practice ISBN 0 7176 0792 5

U9
A guide to managing health and safety in construction. 1995 ISBN 0 7176 0755 0

Fire safety during construction work. Guidance for clients, designers and those managing and carrying out construction work. (In preparation, 1996)

Chapter 7

Major hazards and building evaluation

HS(G)25
Control of Industrial Major Accident Hazards Regulations 1984 (CIMAH): further guidance on emergency plans. 1985 ISBN 0 11 883831 8

HS(R)21 (Rev)
A guide to the Control of Industrial Major Accident Hazards Regulations 1984. 1990 ISBN 0 11 885579 4

HS(G)58
Evaluation and inspection of buildings and structures. 1990 ISBN 0 11 885441 0

GS 51
Facade retention. 1992 ISBN 0 11 885727 4

Chapter 10

Personal protective equipment

L25
Personal Protective Equipment at Work Regulations 1992. Guidance on regulations. 1992 ISBN 0 7176 0415 2
Construction (Head Protection) Regulations 1989. 1990 ISBN 0 11 885503 4

HS(G)66
Protection of workers and the general public during the development of contaminated land. 1991 ISBN 0 11 885657 X

HS(G)53
Respiratory protective equipment: a practical guide for users. 1990 ISBN 0 11 885522 0

GS 27
Protection against electric shock. 1984 ISBN 0 11 883583 1

Chapter 11

Accidents and first aid

COP 42
First aid at work. Health and Safety (First Aid) Regulations 1981. Approved code of practice and Guidance. 1990 ISBN 0 7176 1012 8

L73
Guide to the Reporting of Injuries, Diseases and Dangerous Occurrences Regulations 1995. 1996 ISBN 0 7176 1012 8

Combined pad of F2508 and F2508A forms for reporting injuries, diseases or dangerous occurrences. ISBN 0 7176 1078 0

HSE Books

HSE books may be obtained from HSE Books, PO Box 1999, Sudbury, Suffolk CO10 6FS Tel 01787 881165 Fax 01787 313995

Appendix 5
Dangerous occurrences to be reported (abridged)

A complete list of reportable dangerous occurrences is in the pad of report forms (ISBN 0 7176 1078 0) obtainable from HSE Books and in Appendix 2 of RIDDOR 95.

Reportable dangerous occurrences are

1. collapse, overturning or failure of load-bearing parts of lifts and lifting equipment

2. explosion, collapse or bursting of any closed vessel or associated pipework

3. failure of any freight container in any part of its load-bearing parts

4. plant or equipment coming into contact with overhead power lines

5. electrical short circuit or overload causing fire or explosion

6. any unintentional explosion, misfire, failure of demolition to cause the intended collapse, projection of material beyond a site boundary, injury caused by an explosion

7. accidental release of a biological agent likely to cause severe human illness

8. failure of industrial radiography or irradiation equipment to de-energise or return to its safe position after the intended exposure period

9. malfunction of breathing apparatus while in use or during testing immediately before use

10. failure or endangering of diving equipment, the trapping of a diver, an explosion near a diver, or an uncontrolled ascent

11. collapse or partial collapse of a scaffold over 5 m high, or erected near water where there could be a risk of drowning after a fall

12. unintended collision of a train with any vehicle

13. dangerous occurrence at a well (other than a water well)

14. dangerous occurrence at a pipeline

15. failure of any load-bearing fairground equipment, or derailment or unintended collision of cars or trains

16. a road tanker carrying a dangerous substance overturns, suffers serious damage, catches fire or the substance is released

17. a dangerous substance being conveyed by road is involved in a fire or released

The following dangerous occurrences are reportable except in relation to offshore workplaces

18. unintended collapse of: any building or structure under construction, alteration or demolition where over 5 t of material falls; a wall or floor in a place of work; any falsework

19. explosion or fire causing suspension of normal work for over 24 hours

20. sudden, uncontrolled release in a building of: 100 kg or more of flammable liquid: 10 kg of flammable liquid above its boiling point; 10 kg or more of flammable gas; or of 500 kg of these substances if the release is in the open air

21. accidental release of any substance which may damage health

Note: Additional categories of dangerous occurrences apply to mines, quarries, relevant transport systems (railways etc) and offshore workplaces.

Index